A Student Handbook for Writing in Biology

Fourth Edition

A STUDENT HANDBOOK FOR WRITING IN BIOLOGY

FOURTH EDITION

Karin Knisely
Bucknell University

SINAUER ASSOCIATES, INC.

MACMILLAN

Cover photos

Front cover: flower, *Protea* sp.; leopard, *Panthera pardus*; Great Frigatebird, *Fregata minor*. Back cover: elephants, *Loxodonta* sp.; meerkats, *Suricata suricatta*; cycad, *Encephalartos* sp.; chameleon, species unknown. Photos by Andrew Sinauer.

A Student Handbook for Writing in Biology, Fourth Edition
Copyright © 2013 by Sinauer Associates, Inc. All rights reserved. This book may not be reproduced in whole or in part without permission.

Address editorial correspondence to:
Sinauer Associates, Inc., 23 Plumtree Road, Sunderland, Massachusetts 01375 U.S.A.
www.sinauer.com
publish@sinauer.com

Address orders to:
MPS/W. H. Freeman & Co. Order Department, 16365 James Madison Highway, U.S. Route 15, Gordonsville, VA 22942 U.S.A.

Library of Congress Cataloging-in-Publication Data

Knisely, Karin.
  A student handbook for writing in biology / Karin Knisely, Bucknell University. -- Fourth edition.
        pages cm
  Includes bibliographical references and index.
  ISBN 978-1-4641-5076-0
  1. Biology--Authorship--Style manuals.   I. Title.
  QH304.K59 2013
  808'.06657--dc23

                                                        2012047296

Printed in U.S.A.
5 4

To those who go the extra mile in pursuit of excellence

CONTENTS

PREFACE

The goal of this handbook is to provide students with a practical, readable resource for communicating their scientific knowledge according to the conventions in biology. Chapter 1 introduces the scientific method and experimental design. Because the scientific method relies heavily on the work of other scientists, students must learn how to locate primary references efficiently, quite often using article databases and scholarly search engines (Chapter 2). Once a suitable journal article has been found, the next challenge is to read and understand the content. Chapter 3 describes scientific paper tone and format, provides strategies for reading technical material, emphasizes the importance of paraphrasing when taking notes, and gives examples of how to present and cite information to avoid plagiarism. Using the standards of journal publication as a model, students are then given specific instructions for writing their own laboratory reports with accepted format and content, self-evaluating drafts, and using peer and instructor feedback to refine their writing (Chapters 4 and 5). Besides writing about it, scientists communicate scientific knowledge through posters and oral presentations. How these presentation forms differ from papers in terms of purpose, content, and delivery is the subject of Chapters 7 and 8.

Scientific communication requires more than excellent writing skills—it requires technical competence on the computer. Most first-year students have had little experience producing Greek letters and mathematical symbols, sub- and superscripted characters, graphs, tables, drawings, and equations. Yet these are characteristics of scientific papers that require a familiarity with the computer beyond basic keyboarding skills. For exactly this reason, the appendices, which make up almost half of the book, are devoted to Microsoft Word, Excel, and PowerPoint features that enable scientists to produce professional quality papers, graphs, posters, and oral presentations effectively and efficiently. In the Fourth Edition, the appendices have been updated for Microsoft Office 2010 running on the Windows 7 operating system. The Word appendix (Appendix 1) has an updated section on online backup options, recognizing our increasing reliance on the Internet for storing files and collaborating with others. The Excel appendix (Appendix 2) now also includes instructions for making graphs in Excel for Mac 2011. A section on saving formats as chart templates has been added to allow you to standardize your graphs and reduce the number of repetitive formatting steps. Various small changes were made to the PowerPoint appendix (Appendix 3) to accompany the changes in the menus. To help users with the transition to Office 2010, each appendix contains a table that summarizes frequently used commands and where they are located in the new programs.

In addition to updates to the Appendices, many of the chapters of the Fourth Edition have been reworked. The information on databases and search engines used to find scholarly information in the biological sciences has been updated in Chapter 2. Instructions for RefWorks are also provided in Chapter 2 to illustrate the usefulness of reference management software. Learning how to use RefWorks reduces the tediousness of organizing and formatting citations and end references in scientific papers. In Chapter 4, the step-by-step instructions for writing a lab report have been restructured to emphasize areas in which undergraduate students typically make mistakes: tense, voice, level of detail, and overall organization. Examples and explanations illustrate how to revise faulty writing. Chapter 5 has also been restructured to provide a more systematic, broad-to-specific approach to revision. This chapter provides guidelines for evaluating overall content and organization, and includes specific tactics to improve writing so that it is both clear and grammatically sound. Checklists in Chapters 4, 5, and 6 summarize the most important areas where revision may be needed.

While some users of this book may enjoy reading it cover-to-cover, the majority will use it primarily as a look-up reference. Most of the sections are designed to stand alone so that readers can look up a topic in the index and find the answer to their question. Those who want to learn more about the topic have the option of reading related sections or entire chapters. Tabbed pages in Chapters 1–8 make it easy to find reference formats and the Biology Lab Report and Laboratory Report Mistakes checklists. The Fourth Edition is also available in an e-version for those who prefer to "pack light" and search rapidly and efficiently.

The book is augmented by ancillary materials available on the Sinauer Associates website. The Biology Lab Report Template, as a Microsoft Word document, provides prompts that help students get used to scientific paper format and content. The Biology Lab Report Checklist can be printed out to help students self-evaluate or peer review lab reports. Both instructors and students may find the Laboratory Report Mistakes table handy for use in both the revision and feedback stages. The Evaluation Form for Oral Presentations enables listeners to provide feedback to the speaker on things that he/she is doing well as well as areas that need improvement. Similarly, the Evaluation Form for Poster Presentations can be used as a checklist for the presenter and an evaluation tool for participants during the actual poster session. To illustrate principles of designing effective posters, sample posters are posted on the Sinauer Associates website ⟨<http://sites.sinauer.com/Knisely4E>⟩ and each poster is accompanied by a short evaluation of the layout and content.

Acknowledgments

Feedback from users of previous editions guided changes made in the Fourth Edition. In particular, I would like to thank my students in Introduction to Molecules and Cells (BIOL205) and Organismal Biology (BIOL206) laboratories for giving me suggestions on how to improve this book. The following comment exemplifies the attitude of many first- and second-year biology students toward writing, and makes me smile: "I don't enjoy writing lab reports, but your book was extremely helpful in doing so." My colleagues in the Biology Department at Bucknell University continually provide me with fresh and refreshing ideas on teaching and I am proud to be part of such a collegial group. I am indebted to the helpful professionals in Library and Information Technology at Bucknell for answering my research and computer-related questions, especially Kathleen McQuiston (article databases and scholarly search engines), Dave Kline and Melissa Rycroft (installing the PatternUI add-in and making chart templates in Excel), Jim Van Fleet (RefWorks and research), Debra Balducci (photographing and editing posters), and Bud Hiller (options for backing up files). Special thanks go to Carmen Acuna (Bucknell University) for all the good things I learned in her statistics class, to Marvin H. O'Neal III (Stony Brook University) for his suggestion to add a section on formulas and for making me aware of Microsoft's online interactive guide, and to Teresa Bell, a SkillPath Seminars presenter, for her many practical suggestions on "How to Build Powerful PowerPoint Presentations." I would also like to thank Sandy Field and Kathy Shellenberger for providing feedback on Chapter 2 and for many enjoyable conversations. Kathy also made the screenshots in Excel for Mac 2011 and provided technical assistance with the Mac commands. Finally, I would like to thank my children for keeping me up-to-date on the latest technological developments. My son, Brian Knisely, especially has always been willing to share his technical knowledge with me, communicating concepts with empathy and clarity. He also made the Word 2010 and PowerPoint 2010 screenshots in the Fourth Edition.

I also express my sincere appreciation to those who contributed to the previous editions: John Woolsey (poster presentations), Lynne Waldman (good student lab report), Joe Parsons (University of Victoria), Randy Wayne (Cornell University), John Byram (W.W. Norton & Co.), Kristy Sprott (The Format Group LLC), Wendy Sera (Baylor University), Elizabeth Bowdan (University of Massachusetts at Amherst), Mary Been (Clovis Community College), and the staff of the Bucknell Writing Center. Special thanks also to Walter H. Piper (Chapman University) for his thoughtful comments on the use of passive and active voice, Warren Abrahamson (Bucknell University) for helping me clarify the use of trendlines, and Karen Patrias (National Library of Medicine) for her insight on citation and reference styles.

I am grateful to all the professionals at Sinauer Associates and W.H. Freeman who helped with the production of this book, especially Andy Sinauer, who provided encouragement, feedback, and organizational input on all four editions. I would also like to thank Dean Scudder for starting the conversation on the new editions and for help with various requests along the way. Azelie Aquadro, Christopher Small, and Jefferson Johnson guided the Fourth Edition through the editorial and production processes. Jefferson deserves the credit for the cool cover and Jason Dirks and Nate Nolet set up the website for the ancillary materials.

My passions outside of teaching biology helped shape the style and content of this book. I have always loved sports and, for the past 13 years, have helped coach girls' lacrosse teams in the Central Susquehanna Valley. Just as coaching and teaching go hand in hand, excelling on the playing field and in the classroom both require preparation, discipline, teamwork, and a willingness to do one's best. Through lacrosse, I have had the privilege of meeting players, parents, coaches, and umpires who have helped me become a better teacher and coach.

As a freelance German to English technical translator, I appreciate the challenge of choosing just the right words to communicate information faithfully, in a clear and concise manner, and according to conventions familiar to the reader. I am indebted to the translation profession for setting high standards of quality. While the process of writing in a particular genre is challenging, it is also immensely satisfying: We experience the joy of learning, the pleasure of sharing our knowledge with others, and the creativity involved in putting thoughts into words. Especially rewarding is the enthusiasm we generate when we communicate our knowledge well.

Lastly, I would like to thank my family and friends for their unwavering encouragement and love. My late father, Adolph Wegener, instilled in me a love of learning and supported me in all of my endeavors; I will miss him. My mother, Elfriede, has been a role model and inspiration for me throughout my life. My children, Katrina, Carleton, and Brian, continue to be my pride and joy. I would not be able to pursue all of my passions without the support of my husband, Chuck, whose love and sense of humor sustain me.

KARIN KNISELY
LEWISBURG, PA
DECEMBER 2012

Chapter 1

THE SCIENTIFIC METHOD

Trying to understand natural phenomena is human nature. We are curious about why things happen the way they do, and we expect to be able to understand these events through careful observation and measurement. This process is known as the scientific method, and it is the foundation of all knowledge in the biological sciences.

An Introduction to the Scientific Method

The scientific method involves a number of steps:

- Asking questions
- Looking for sources that might help answer the questions
- Developing possible explanations (hypotheses)
- Designing an experiment to test a hypothesis
- Predicting what the outcome of an experiment will be if the hypothesis is correct
- Collecting data
- Analyzing data
- Developing possible explanations for the experimental results
- Revising original hypotheses to take into account new findings
- Designing new experiments to test the new hypotheses (or other experiments to provide further support for old hypotheses)
- Sharing findings with other scientists

Most scientists do not rigidly adhere to this sequence of steps, but it provides a useful starting point for how to conduct a scientific investigation.

Ask a question

As a biology student, you are probably naturally curious about your environment. You wonder about the hows and whys of things you observe. To apply the scientific method to your questions, however, the phenomena of interest must be sufficiently well defined. The parameters that describe the phenomena must be measurable and controllable. For example, let's say that you learned that:

> Dwarf pea plants contain a lower concentration of the hormone gibberellic acid than wild-type pea plants of normal height.

You might ask the question:

> Does gibberellic acid regulate plant height?

This is a question that can be answered using the scientific method, because the parameters can be controlled and measured. On the other hand, the following question could not be answered easily with the scientific method:

> Will the addition of gibberellic acid increase a plant's sense of well-being?

In this example, "a sense of well-being" is not something that can be measured or controlled.

Look for answers to your question

There is a good chance that other people have already asked the same question. That means that there is a good chance that you may be able to find the answer to your question, if you know where to look. Secondary references such as your textbook, encyclopedias, and information posted on the websites of university research groups, professional societies, museums, and government agencies are usually easier to comprehend than journal articles and may be good places to begin finding answers (see the section "Understand your topic" in Chapter 2). Curiously, attempts to answer the original question often result in new questions, and unexpected findings lead to new directions in research. By reading other people's work, you may think of a more interesting question, define your question more clearly, or modify your question in some other way.

Turn your question into a hypothesis

As a result of your literature search or conversations with experts, you may now have a tentative answer to your original (or modified) question. Now it is time to develop a hypothesis. A **hypothesis** is a possible explanation for something you have observed. **You must have information before you can propose a hypothesis!** Without information, your hypothesis is nothing more than an uneducated guess. That is why you must look for possible answers before you can turn your question into a hypothesis.

A useful hypothesis is one that can be tested and either supported or negated. A hypothesis can never be *proven* right, but the evidence gained from your observations and/or measurements can *provide support for* the hypothesis. Thus, when scientists write papers, they never say, "The results prove that… ." Instead, they write, "The results suggest that…" or "The results provide support for… ."

You might transform your question "Does gibberellic acid regulate plant height?" into the following testable hypothesis:

> Good: The addition of gibberellic acid to dwarf plants will allow them to grow to the height of normal, wild-type plants.

This hypothesis provides specific expectations that can be tested. In contrast, the following hypothesis is not specific enough:

> Vague: The addition of gibberellic acid will affect the height of dwarf plants.

Design an experiment to test your hypothesis

In an **observational study**, scientists observe individuals and measure variables of interest without trying to control the variables or influence the response. While observations provide important information about a group, it is difficult to draw conclusions about cause and effect relationships because multiple factors affect the response. That's the main reason why scientists conduct experiments. **Experiments** are studies in which the investigator imposes a specific treatment on a person or thing while controlling the other factors that might influence the response.

The first step in designing an experiment is to determine which variables might be influential. Of those variables, only one may be manipulated in any given experiment; the others have to remain constant. The individuals in the experiment are then divided into treatment and control groups. The treatment group is subjected to the independent variable and the control group is not; all other conditions are the same for the two groups. If the hypothesis is supported, the individuals in the treatment group will respond differently from those in the control group. If there is no difference in response between the treatment and control groups, the so-called null hypothesis is supported. Having enough replicates lends assurance that the results are reliable.

Define the variables.   Variables are commonly classified as independent or explanatory variables, dependent or response variables, and controlled variables. The *one* variable that a scientist manipulates in a given experiment is called the **independent variable** or the explanatory variable, so-called because it "explains" or influences the response. It is important to manipulate *only one* variable at a time to determine whether or not a cause and effect relationship exists between that variable and an individual's response. The other variables that may affect the response must be carefully controlled so that they do not confound the relationship between the independent variable and the dependent variables.

  Dependent variables are those affected by the imposed treatment; in other words, they represent an individual's response to the independent variable. Dependent variables are variables such as size, number of seeds produced, and velocity of an enzymatic reaction, which can be measured or observed.

  The hypothesis proposed earlier involves testing whether there is a cause and effect relationship between gibberellic acid (GA) treatment and plant height. GA level is the variable that will be manipulated; plant height is the response that we'll measure. Because plant height is affected by many other factors such as ambient temperature, humidity, age of the plants, day length, amount of fertilizer, and watering regime, however, we must keep these controlled variables constant so that any differences in response can be attributed to the GA treatment.

Set up the treatment and control groups.   The individuals in the experiment are assigned randomly to either a treatment group or a control group. Those in the treatment group will be subjected to the independent variable, while those in the control group will not. Depending on the hypothesis, the control group may be subdivided into positive and negative controls. Negative controls are not treated with the independent variable and are not expected to show a response. Positive controls are individuals not treated with the independent variable, but represent a reference for treatment groups that demonstrate a response consistent with the hypothesis.

Hypothesis:	Adding GA to dwarf plants will allow them to grow to the height of normal, wild-type plants.
Treatment group:	Dwarf plants + GA
Control groups:	
Negative:	Dwarf plants + no GA (substitute an equal volume of water)
Positive:	Wild-type plants + no GA

Determine the level of treatment for the independent variable. How much GA should be added to the dwarf plants in the treatment group to produce an increase in height? Too little GA may not effect a response, but too much might be toxic. To determine the appropriate level of treatment, consult the literature or carry out a preliminary experiment. The level may even be a range of concentrations that is appropriate for the biological system.

Provide enough replicates. A single result is not statistically valid. The same treatment must be applied to many individuals and the experiment must be repeated several times to be confident that the results are reliable.

Make predictions about the outcome of your experiment. Predictions provide a sense of direction during both the design stage and the data analysis stage of your experiment. For each treatment and control group, predict the outcome of the experiment if your hypothesis is supported. You may also choose to propose a **null hypothesis**, which states that the treatment has no effect on the response.

Hypothesis:	Adding GA to dwarf plants will allow them to grow to the height of normal, wild-type plants.
Treatment group:	Dwarf plants + GA
Prediction if hypothesis is supported:	Dwarf plants will grow as tall as wild-type plants + no GA.
Null hypothesis:	Dwarf plants will not grow to the height of wild-type plants.
Negative control:	Dwarf plants + no GA
Prediction:	Dwarf plants will be short.
Positive control:	Wild-type plants + no GA
Prediction:	Wild-type plants will be tall.

Record data

Scientists record procedures and results in a laboratory notebook. The type of notebook (bound or loose leaf, with or without duplicate pages) may be prescribed by your instructor or the principal investigator of the research lab. More important than the physical notebook, however, is the detail and accuracy of what's recorded inside. For each experiment or study, include the following information:

- Investigator's name
- The date (month, day, and year)
- The purpose
- The procedure (in words or as a flow chart)
- Numerical data, along with units of measurement, recorded in well-organized tables
- Drawings with dimensions and magnification, where appropriate. Structures are drawn in proportion to the whole. Parts are labeled. Observations about the appearance, color, texture, and so on are included.
- Graphs, printouts, and gel images
- Calculations
- A brief summary of the results
- Questions, possible errors, and other notes

When deciding on the level of detail, imagine that, years from now, you or someone else wants to repeat the experiment and confirm the results. The more information you provide, the easier it will be to understand what you did, what problems you encountered, suggestions for improving the procedure, the results you obtained, how you summarized the data, and how you reached your conclusions.

Summarize numerical data

The raw data in lab notebooks are the basis for the results published in the primary and secondary literature. Published results, however, usually represent a *summary* of the raw data by the author, who is both knowledgeable about the subject and intimately familiar with the experiment. We rely on the author's experience and integrity to reduce the original data to a more manageable form that is an honest representation of the phenomenon and which lends itself to interpretation.

How the author presents data in the Results section depends in part on the scope of the question asked at the beginning. Broad questions about a population involve **statistical inference**, whereby results from a sample or subset of the population are applied to the whole. Because a different sample may produce different results, the author includes a statement about the reliability of his or her conclusions using appropriate statistical language. On the other hand, narrower questions about a specific situation may be answered from the data at hand. For example, questions such as "Which fraction of a purification procedure contains the most enzymatic activity?" or "Which medium produces the highest concentration of bacteria?" can be answered from the collected data and require no inference about a larger population. When the data are consistent from one experiment to the next, scientists gain confidence that their conclusions are valid.

When *you* are given the task of summarizing the raw data, first distinguish between trustworthy and erroneous data. Erroneous data include results obtained by dubious means, for example, by not following the procedure, using the equipment improperly, or making simple arithmetic errors. Trustworthy data include results obtained legitimately, but which may still have quite a bit of unexplained variability. If time permits, repeat the experiment to determine possible sources of variability and make changes in the procedure if necessary.

Once you've identified which data are reliable, graph them. It is easier to spot patterns and outliers on a graph than in a table. Furthermore, graphs are used to check assumptions for certain statistical methods. Use bar graphs when one of the variables is categorical (i.e., it has no units of measurement). Use scatterplots and line graphs when both variables are quantitative. Look for an overall trend as well as deviations from the trend. Reduce the data by taking the average (mean) and express variability, where appropriate, in terms of standard deviation or standard error. Never eliminate data without a good reason.

Analyze the data

Once you have a visual summary of the raw data, look for relationships between variables. Do the results match the predictions if the hypothesis is supported? If so, then compare your results to those in the primary references you consulted to develop your hypothesis in the first place. Comparable data from different studies help researchers gain assurance that their conclusions about a particular phenomenon are valid. When analyzing data, however, do not let your predictions affect your objectivity. Do not make your results fit your predictions—instead, modify your hypothesis to fit your results. What is learned from a negated hypothesis can be just as valuable as what is learned from a "successful" experiment.

Keep in mind that there may be no difference between the control and the experimental treatments. If there was no difference, say so, and then try to develop possible explanations for these results.

Try to explain the results

Once you have summarized and analyzed the data, you are ready to develop possible explanations for the results. You previously found information on your topic when you developed your hypothesis. Return to this material to try to explain your results. Do your results agree with those of other researchers? Do you agree with their conclusions? If your results do not agree, try to determine why not. Were different methods, organisms, or conditions employed? What were some possible sources of error?

You should realize that even some of the most elementary questions in biology have taken hundreds of scientists many years to answer. One approach to the problem may seem promising at first, but as data are collected, problems with the method or other complications may become apparent. Although the scientific method is indeed methodical, it also requires imagination and creativity. Successful scientists are not discouraged when their initial hypotheses are discredited. Instead, they are already revising their hypotheses in light of recent discoveries and planning their next experiment. You will not usually get instant gratification from applying the scientific method to a question, but you are sure to be rewarded with unexpected findings, increased patience, and a greater appreciation for the complexity of biological phenomena.

Revise original hypotheses to take new findings into account

If the data support the hypothesis, then you might design additional experiments to strengthen the hypothesis. If the data do not support the hypothesis, then suggest modifications to the hypothesis or use a different procedure. Ideally, scientists will thoroughly investigate a question until they are satisfied that they can explain the phenomenon of interest.

Share findings with other scientists

The final phase of the scientific method is communicating your results to other scientists, either at scientific meetings or through a publication in a journal. When you submit a paper to refereed journals, it is read critically by other scientists in your field, and your methods, results, and conclusions are scrutinized. If any errors are discovered, they are corrected before your results are communicated to the scientific community at large.

Poster sessions are an excellent way to share preliminary findings with your colleagues. The emphasis in poster presentations is on the methods and the results. The informal atmosphere promotes the exchange of ideas among scientists with common interests. See Chapter 7 on how to prepare a poster.

Oral presentations are different from both journal articles and poster sessions, because the speaker's delivery plays a critical role in the success of the communication. See Chapter 8 for tips on preparing and delivering an effective oral presentation.

Chapter 2

Developing a Literature Search Strategy

The development of library research skills is an essential part of your training as a biology student. A vast body of literature is available on just about every topic. Finding exactly what you need is the hard part.

In biology, sources are divided broadly into primary and secondary references. **Primary references** are the research articles, dissertations, technical reports, or conference papers in which a scientist describes his or her original work. Primary references are written for fellow scientists—in other words, for a specialized audience. The objective of a primary reference is to present the essence of a scientist's work in a way that permits readers to duplicate the work for their own purposes and to refute or build on that work.

Secondary references include encyclopedias, textbooks, articles in popular magazines, and information posted on the websites of professional societies, government agencies, and other scientific organizations. Secondary references are based on primary references, but they address a wider, less-specialized audience. In secondary references, there is less emphasis on the methodology and presentation of data. Instead, the results and their implications are described in general terms for the benefit of non-specialist readers.

You will delve into the biological literature when you write laboratory reports, research papers, and other assignments. Although secondary references provide a good starting point for your work, it is important to be able to locate the primary sources on which the secondary sources are based. Only the primary literature provides you with a description of the methodology and the actual experimental results. With this information, you can draw your own conclusions from the author's data.

Although initially it may be difficult to read primary literature, it will become easier with practice, and the rewards are well worth it. One benefit of *reading* research articles is that you will become a better *writer*. Through reading, you become familiar with the writing style and overall structure of research articles, so that you have a model when you write your own lab reports. Another benefit is that you learn how scientists approach a problem, design experiments to test hypotheses, and interpret their results to arrive at their conclusions. Emulating their writing style may help you improve your critical thinking skills. A further benefit of reading the primary literature is getting to know the scientists who work in a particular subdiscipline. You may discover that you are sufficiently interested in a subdiscipline to pursue graduate work with one or more of the authors of a journal article.

How do you find primary references that are directly relevant to your topic? The fastest and easiest way is to search article databases. **Article databases** contain a pre-screened collection of scholarly information, not web

TABLE 2.1 Databases and search engines used to find scholarly information in the biological sciences

Database or Search Engine	Description
AGRICOLA	Produced by the National Agricultural Library, this database contains citations for journal articles, monographs, government publications, and other types of publications in the field of agriculture and related areas.
Biological Abstracts	Considered the most comprehensive database in the area of biology and the life sciences, it provides abstracts and citations to journal literature.
Biological Sciences (ProQuest)	Contains abstracts and citations to journals, monographs, and other publication types for a wide range of areas in the life sciences.
BioOne	Contains the full text of peer-reviewed articles in the biosciences. Most of the journals are published by small scientific societies, other not-for profits, and open access publishers.
Google Scholar	A Web search engine for scholarly literature across many disciplines. Includes not only journal articles, but also material from websites of universities, scientific research groups, and professional societies; conference proceedings; and preprint archives (preprints are manuscripts circulated because they contain current information, but they have not yet been peer reviewed).
JSTOR	Developed as an archive of core scholarly journals, this database searches the full text of core journals in a variety of disciplines including biology and ecology. Coverage begins with the first issue of each journal. However there is a gap, typically from 1 to 5 years, between the most recently published issue and when it appears in JSTOR.

pages that anyone could have created. Article databases are owned by companies or organizations that employ experts to read scholarly articles and then enter information about the articles into the database. To find scholarly information on a particular topic, instead of "googling" the entire Web, you will typically search one or more databases.

Most databases (PubMed being the notable exception) are by subscription. Companies that own these databases sell licensing agreements to university libraries and other institutions. If you are affiliated with such a university or institution, then you can use fee-based databases for free. On the other hand, search engines such as Google Scholar and Scirus, which scan the Web for scientific information, are free and available to the general public. Table 2.1 describes some of the databases and scholarly search engines that you may have access to.

TABLE 2.1 *Continued*	
Database or Search Engine	**Description**
NCBI (National Center for Biotechnology Information)	A division of the National Library of Medicine. Produces searchable databases on nucleotide and protein sequences, protein structures, complete genomes, taxonomy, and other molecular biology information.
PubMed	Produced by the National Library of Medicine, PubMed is the public access version of MEDLINE, the premier database for medicine and related fields. It contains abstracts and citations to the worldwide journal literature.
Science Direct	Provides access to journal articles and books published by Elsevier. Although multidisciplinary, most references are in the areas of science, medicine, and engineering.
Scirus—for scientific information only	A Web search engine and database of scientific information resources, covering websites to journal articles. Like Google Scholar, this search engine is free and some of the content may not be peer-reviewed.
Scopus	The world's largest abstract and citation database of peer-reviewed literature across the scientific, technical, and medical fields, the social sciences, and arts and humanities.
Web of Science	An interdisciplinary database for peer-reviewed articles from core journals in many subject areas. The Cited Reference Search allows you to identify articles that cite a particular author or work.

Source: Bucknell University, Library and Information Technology [Internet]. Lewisburg (PA): Bucknell University; c2008 [cited 2012 Oct 28]. Available from: http://www.bucknell.edu/script/ISR/Databases/

Most of this chapter describes how to find primary references using databases. If you do not have access to these databases, however, you can still locate references the old-fashioned way. This method involves building a bibliography from sources cited in books, journal articles, and other literature. Books such as *Annual Reviews* are considered secondary references, but the Literature Cited section in review articles often is an excellent source of primary references. Building a bibliography without the use of a database is laborious and time-consuming, but the end result is often the same. An advantage of using this old-fashioned method is that you may find older, seminal papers that may not be indexed in databases. If your assignment requires a thorough search of the literature, you will most likely use a combination of database and manual searches. Don't forget about your human resources—seek assistance from your reference librarian during all stages of your research project.

Databases and Search Engines for Scientific Information

Familiarize yourself with the databases and search engines recommended by your professor or a reference librarian and which are available through your academic library. All of the databases have some overlap in terms of the journals they index, but there are also unique listings. Results may also vary depending on subject and publication year.

Comparison of databases

One of the great things about electronic databases is that they are continually updated and improved, giving you access to the most current scientific information available on the Internet. But with so many choices and so little time, what's the best strategy for tracking down a few good primary journal articles for your topic? The answer to this question depends on who you ask, how comprehensive your research needs to be, the subject matter, and personal search preferences. Nonetheless, knowing a little about the strengths and weaknesses of some of the major databases and search engines may help guide your strategy (Table 2.2).

Librarians have published a number of recent papers on this topic (see, for example, Falagas and others [2008], Shultz [2007], Bakkalbasi and others [2006], and Giustini and Barsky [2005]). While these published comparisons are as transient as the databases they describe, it is nonetheless instructive to look at some of the data.

Google Scholar. Google Scholar was introduced by Google in 2004. Its strengths are name recognition, a simple query box, and the fact that it's free. In terms of content, Google Scholar is thought to provide greater access to older

TABLE 2.2 Comparison of features of selected biology databases and search engines

	Biological Abstracts	Google Scholar	PubMed	Web of Science	Scopus
Resource type	Database	Search engine	Database	Database	Database
Access (free or fee-based)	Fee-based (usually institutional subscription)	Free	Free	Fee-based	Fee-based
Years covered	1926 to present	Unknown	1950 to present	1900 to present	1823 to present
Sources retrieved	Journal articles	No information provided, but retrieves journal articles, books, preprints, abstracts, technical reports, and other electronic media	Journal articles, literature reviews, clinical trials	Journal articles and conference proceedings	Articles, conference papers
Content (number of journals indexed)	Searches more than 4,200 journals in the life sciences	Unknown	Biomedical journal citations and abstracts from more than 5,000 journals	Searches more than 10,000 journals (all disciplines) and 110,000 conference proceedings	Nearly 18,000 titles from more than 5,000 international publishers from all subject areas
Reliability (peer-reviewed materials)	Most journals are peer-reviewed	Unclear whether all journal articles are peer-reviewed	Most journals are peer-reviewed	All journals are peer-reviewed	All journals are peer-reviewed

Source: Kathleen McQuiston, Research Services Librarian, Library and Information Technology, Bucknell University (2012 Oct 28) and respective database or search engine websites.

records and to material not easily located through conventional channels such as publishers' websites. Some of Google Scholar's weaknesses include the scope of its coverage (it finds too much information), uncertainty about the scholarly value and currency of some of the records, and the sorting of records according to how relevant and popular they are (not how current).

PubMed. PubMed is *the* most recommended database for researchers in medicine who require advanced search functions. Like Google Scholar and Scirus, PubMed is free and its advanced search feature makes it possible to limit searches by author, publication, and date. PubMed provides a variety of options to retrieve only certain formats (full text, free full text, or abstract), types of article (clinical trial, review, clinical conference, comparative study, government publication, etc.), language, and content (journal group, research topic, humans or animals, gender, and age). Another feature that makes PubMed so powerful is its search algorithm, which is based on concept recognition, not letters or words. Every document indexed for PubMed has been read by experts, who tag the document with controlled vocabulary (Medical Subject Headings or MeSH) that accurately describes the paper's content. "False hits" due to homographs (e.g., swimming pool rather than gene pool) are thus eliminated in PubMed searches. Furthermore, MeSH solves the problem of ambiguity concerning scientific and popular names of organisms, synonyms, and variations in British and American spelling.

Web of Science. Web of Science is fee-based, so you may only have access to this database if your university has a subscription. Web of Science covers a larger period of time than either PubMed or Google Scholar. It has depth and scope and is useful for finding information on topics of an interdisciplinary nature. The greatest benefit of this database, however, lies in the fact that once you have found a good journal article, you can expand your bibliography quickly based on common references. With Web of Science, it is possible to search *forward* in time to find more recent papers that have cited the paper of interest. It is also possible to search *backward* to find papers cited by authors of the paper of interest.

Scopus. Like Web of Science, Scopus has a tremendous scope in terms of years covered and sources retrieved, and it is fee-based. Scopus, like all of the databases in Table 2.2, has an advanced search feature, provides links to full-text articles, and allows references to be exported to reference management software (see p. 24). In addition, graduate students and career researchers will find the email alerts feature of these databases handy for staying current with the literature. When registering for email alerts, you can enter keywords that are relevant to your research. When a new article containing these keywords appears, the database administrator will send you an email alert.

Database Search Strategies

Finding just the right journal articles on your topic can be a daunting task. This section will help you get started.

Understand your topic

A productive and efficient search begins with a **basic understanding of your topic**. If you don't even know where to start, look up the most specific term you can come up with in the index of your textbook. Open the book to the pages that contain this term. Read the chapter subheadings and the chapter title to learn how this term fits into the bigger picture. Read the relevant pages to find out what subtopics are associated with this term.

Your library's stacks are another good place to find general information. Search the library's catalog to locate a book on your topic. Write down the call number and find this book on the shelf. Browse the titles of other books in the vicinity. Because the Library of Congress cataloging system groups books according to topic, you can often find additional sources shelved nearby.

Encyclopedias and dictionaries may also help you clarify your topic. Check your library's homepage for references that you may have access to, both electronic and printed sources. Websites such as Wikipedia (http://www.wikipedia.org/), WebMD (http://www.webmd.com), and others may be a good place to start, but evaluate Internet sources critically. Whereas journal articles and books have undergone a rigorous review process, information on the Web may not have been checked by any authority other than the owner of the website.

A first step in evaluating a website's reliability is to look at the ending of the URL address (Table 2.3). Is the sponsor of the website a company or organization that is more interested in trying to sell a product or idea than in presenting factual information? To become a savvy website evaluator, check out the tips on your library's homepage or take one of the tutorials listed in the Bibliography.

Define your research goals

Once you have a basic understanding of your topic, try to **define your research goals** with statements such as

- I would like to compare or contrast methods.
- I'm looking for a cause and effect relationship.
- I want to understand more about a process.

TABLE 2.3 Identifying sponsors of sites on the World Wide Web			
Type of Web Page	Purpose	Ending of URL Address	Examples
Informational	To present (factual) information	.edu, .gov	Dictionaries, directories, information about a topic
Business/marketing	To sell a product	.com	Carolina Biological Supply, Leica
Advocacy	To influence public opinion; to promote the exchange of knowledge and provide resources for its members	.org	Sierra Club, Association for Biology Laboratory Education
News	To present very current information	.com	CNN, *USA Today*
Personal	To present information about an individual	Variety of endings, but has tilde (~) embedded in the URL	

Source: Alexander and Tate (c1996–2005).

- I am interested in how an organism carries out a particular function (e.g., obtains nutrients, reproduces, moves, responds to changes in its environment, etc.).

Subdivide your topic into concepts

Once you have formulated the goals for your topic, start **defining smaller concepts**. For example, if the methods you wish to compare have to do with measuring the amount of protein in a sample, then one of the concepts is protein quantification. Another concept would include the specific names of protein quantification methods, such as Lowry, biuret, Bradford, BCA, and so on. A third concept might relate to the types of protein samples that were analyzed.

Another way to organize concepts related to your topic is to use PubMed's Medical Subject Headings (MeSH) database, a kind of thesaurus for the life sciences. Words entered in the search box are translated into standardized descriptors, which are then listed in a hierarchy of headings and subheadings.

Let's say, for example, that you would like to find concepts related to the topic "How do *Tetrahymena* move?" Go to the PubMed home page

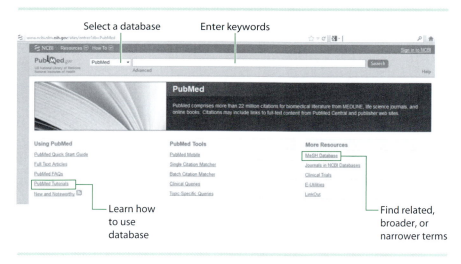

Figure 2.1   PubMed home page provides tutorials and options for searching different databases.

(http://www.ncbi.nlm.nih.gov/sites/entrez?db=PubMed)   and   select
MeSH Database under **More Resources** (Figure 2.1). A search for the term
motility lists *cell movement* as the first result (Figure 2.2); clicking this descriptor opens a page that gives a definition of *cell movement* (not shown in Figure 2.2), entry terms, and the MeSH tree for this concept. The headings below
cell movement in the tree are narrower concepts and the headings above are
broader. Write down the entry terms and headings that are relevant to your
topic. While the entry terms are automatically searched in databases that
use MeSH, they may be useful keyword alternatives in databases or search
engines that do not.

Choose effective keywords

Effective keywords are neither too broad nor too narrow in scope. Keywords
that are too broad will retrieve an unmanageable number of articles that, for
the most part, are not relevant to your topic. On the other hand, keywords
that are too specific may not get any results. For each concept in your topic,
therefore, try to come up with moderately specific terms, synonyms, and
related descriptors (Figure 2.3). Consider different word endings (photosynthetic versus photosynthesis), abbreviations (*HIV* for *human immunodeficiency virus*), and alternative spellings (American versus British English).
Avoid vague terms like *effect* and *relationship between*.

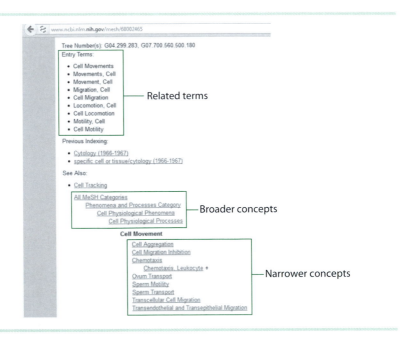

Figure 2.2 MeSH tree for the concept *cell movement*. The Entry Terms shown above the tree are automatically included in a search for the phrase *cell movement* in databases such as PubMed, which use MeSH.

Connect keywords with the operators and, or, or not

After you have generated a list of keywords, select two or more and combine them in a search string using operators such as *and*, *or*, or *not*.

- When the word *and* is used between keywords, the references must have both words present. This connector is a good way to limit your search.

- When the word *or* is used, the references must have at least one of the search terms. This connector is a good way to expand your

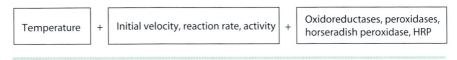

Figure 2.3 Possible keywords generated from concepts related to the topic "How temperature affects the initial velocity of peroxidase, an enzyme isolated from horseradish."

search. For example, the search string *biuret or Bradford* turned up over 7,650 hits in PubMed, while *biuret and Bradford* resulted in only 15 (another search may not result in the same numbers, but the difference would likely be of the same magnitude).

■ When the word *not* is used, then the references should not contain that particular keyword. This connector is another way to limit your search.

Use truncation symbols for multiple word endings

Truncation is a method for expanding your search when keywords have multiple endings. For example, many words related to the concept of temperature begin with *therm*, such as thermoregulation, thermoregulatory, thermy, and thermal. Rather than writing a lengthy search string containing all of these terms, simply type *therm* followed by a wildcard symbol like *, ?, or #. The appropriate truncation symbol can be found in the Help menu of the database you are searching. Google Scholar uses stemming technology instead of truncation, whereby it automatically searches for variations in word endings for the given keyword.

Search exact phrase

When the keyword is a phrase, the search engine typically searches for adjacent words in order. Unfortunately, the search results may also include "false hits" in which the words are separated. When it's important to search an exact phrase, use quotation marks. For example, type *"RNA polymerase"* instead of *RNA polymerase*.

Use the same keywords in a different database or search engine

If you are not having any success with different keyword combinations in one search engine or database, try a different one. Google Scholar and Scirus search the entire Web and may find links to published journal articles on scientists' homepages or course websites. These media are not included in PubMed or Web of Science database searches. Take advantage of the resources at your disposal. Remember that once you find that one good journal article, it will be much easier to find others (see the section "Finding related articles").

Evaluating Search Results

After you type a keyword string into the search box, the search engine goes to work. The result is a page that lists the records by publication date (most

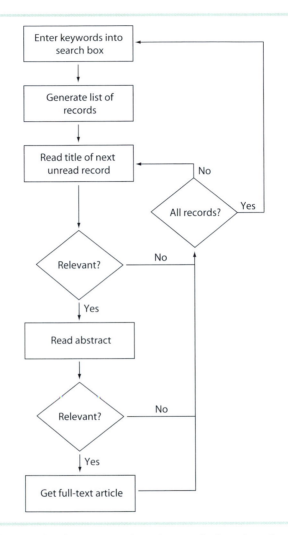

Figure 2.4 Evaluating database or search engine results is an iterative process.

recent first), relevance, or another criterion. Each journal article record contains the article title, the authors' names, the name of the journal, the volume and issue numbers, the pages, and the publication date. Based on the title, decide if you want to read the abstract. After having read the abstract, decide whether you want to read the entire paper. This iterative process is summarized in Figure 2.4.

The results pages for Web of Science and PubMed are formatted slightly differently, but both contain the same basic information about the journal

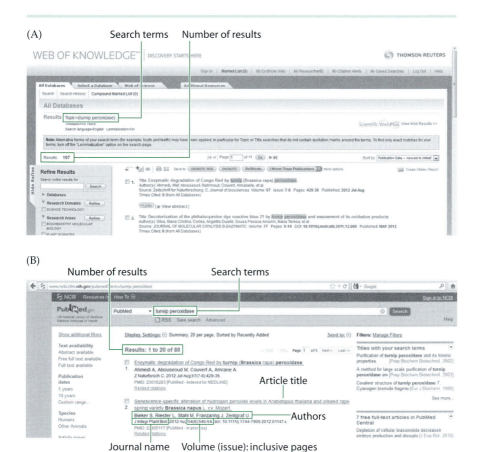

Figure 2.5 The results page from (A) Web of Science and (B) PubMed for the keyword phrase *turnip peroxidase.*

articles (Figure 2.5). You will need this information when you cite the article in your lab report or research paper (see the section "Documenting Sources" in Chapter 4).

Peruse the titles of the first 20 records. If the titles seem to be unrelated to your topic, start a new search with different keywords using the strategies described previously (see the section "Choose effective keywords"). If a title seems promising, click it to open a page that contains the abstract (Figure 2.6). Based on the title and the abstract, decide whether or not you want to read the entire article.

(A) Full-text link

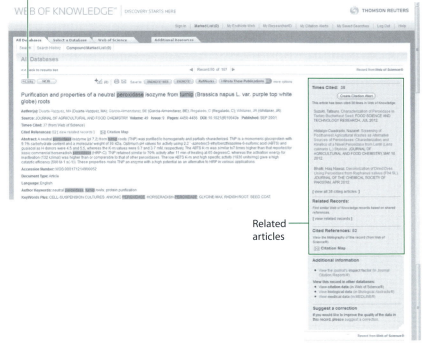

Related articles

Full-text link

(B)

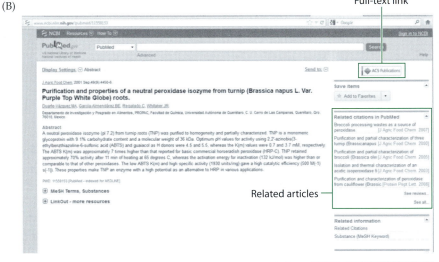

Related articles

Figure 2.6 Detailed record from (A) Web of Science and (B) PubMed showing the abstract, a link to the full-text article, and links to related articles.

Finding related articles

Once you have found a good article, Web of Science makes it easy to find related articles. In the **Times Cited** section, there is a list of more recent papers that cite this article (see Figure 2.6A). Clicking on one of these titles opens a new page that displays the abstract of the more recent paper. In the **Related Records** section, papers are listed, which cite references that were also cited in the article. Common references indicate that the authors were pursuing a similar research topic. In the **Cited References** section, you can view the references listed in the article. Browsing the list allows you to find related papers with a slightly different focus. PubMed also offers a **Related Articles** option (see Figure 2.6B).

Finding **review articles** is the equivalent of hitting the mother lode. Review articles are secondary references that summarize the findings of all major journal articles on a specific topic since the last review. You can find background information, the state of current knowledge, and a list of the primary journal articles authored by scientists who are working on this topic. If you are unable to find a relevant review article in a database, go directly to the Annual Reviews website (http://www.annualreviews.org/) and search for your topic. If you find a promising review article on this website, you may be able to obtain a copy through your academic library.

Most of the article databases and search engines also have an **advanced search** option. Advanced search makes it possible for you to limit your search by specifying one or more authors, publication years, journals, and other criteria.

Obtaining full-text articles

If the title and the abstract of an article sound promising, you will want to obtain the full-text article. Both the Web of Science and PubMed abstract pages have links to full-text articles. To access the full-text article in Web of Science, click the –>**Links** button on the left side above the article title; in PubMed, the link is found at the top right above the **Related Articles** section (see Figure 2.6).

Both databases lead you to the journal publisher's homepage, where you can download the full-text article as a PDF or HTML file (Figure 2.7). PDFs preserve formatting, while HTML files contain hyperlinks that make it easy to access other references. Save the full-text article to your computer or virtual storage space to read later. Copy the URL and write down the download date, because you may need this information when citing the source.

While the abstract is usually free, some publishers charge a fee to access the full-text article. Fortunately, academic libraries and institutions often purchase subscriptions so that faculty, staff, and students can obtain many

Copy URL for later documentation

Figure 2.7 Full-text articles are typically accessed through links to the journal publisher's website. This article can be downloaded as a PDF or HTML file. The full reference can also be downloaded to a citation manager.

electronic journal articles for free. If your library does not have a subscription and you are not in a hurry to get the article, you may be able to use interlibrary loan. **Interlibrary loan** is a way for a library to borrow or obtain materials that it does not own from another library or organization.

Managing References (Citations)

Reference management software makes it possible to

- Build your own collection of references from database searches
- Insert citations into a paper
- Format both the in-text citation and the end reference according to Council of Science Editors style. You can select other styles such as MLA or Chicago Style for papers you write in the humanities.

Some of these products are commercially available (e.g., ProCite, EndNote, and Reference Manager) and others are free as long as you are affiliated with a subscribing institution (e.g., RefWorks and EndNote Web).

Many scientists and other scholars rely heavily on reference management software to organize all of their references. Students will appreciate the convenience and ease of use of these programs as well. The following instructions for RefWorks are intended simply to make you aware of the possibilities. If you like what you see, ask your librarian if you can access something similar at your school.

RefWorks

Create an account. Go to the RefWorks login page found at https://www. refworks.com/refworks2/?r=authentication::init. Enter your institution's Group Code to login. Then create an individual account by clicking **Sign up for an Individual Account** and entering your personal information. Back on the RefWorks Login Center page, log in.

Download citation from database. Using the citation in Figure 2.7 as an example:

1. Click the **Download Citation** link.
2. Click the desired content format: **Citation only**, **Citation and references**, or **Citation and abstract** (Figure 2.8). Click **Download Citation(s)**. The citation file may be downloaded to your computer in RIS format (see next step) or imported directly into the **Last Imported Folder** in RefWorks (skip to Step 4).

Figure 2.8 The **Download Citations** dialog box is used to specify the type of download and the name of the reference management software.

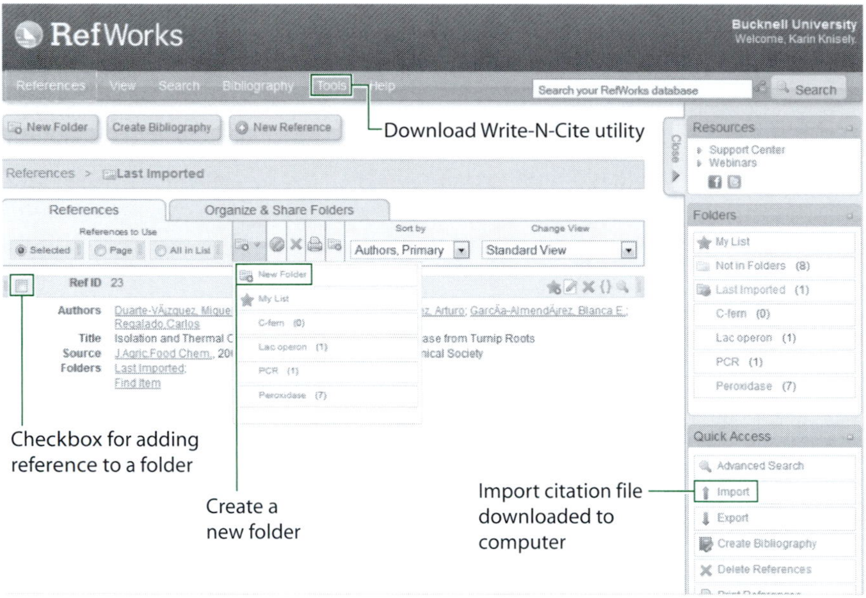

Figure 2.9 Downloaded references can be added to folders to facilitate reference management.

3. In RefWorks, click **Import** and browse your computer for the downloaded citation file. Unless you specify a different folder, the citation will be imported into the **Last Imported Folder**.

4. The RefWorks **Import References** dialog box will notify you that 1 reference was imported. Click **View Last Imported Folder** to see the details of the reference.

5. Click the checkbox for the imported reference (Figure 2.9). Click **New Folder** and type a name or add the reference to an existing folder. References will be easier to find when they are assigned to folders.

Download Write-N-Cite. Write-N-Cite is a utility that allows you to cite sources saved in RefWorks. To download Write-N-Cite, click **Tools | Write-N-Cite** in RefWorks (see Figure 2.9) and follow the installation instructions. Make sure no MS Word documents are open during the installation. Once Write-N-Cite has been installed, open Word and you will notice that a new tab, **RefWorks**, has been added to the Ribbon, as shown in Figure 2.10.

Create an in-text citation and end reference list. When you write lab reports and research articles, you will cite the work of others and then, at the end of your paper, list the full references. Scientists follow the Council of Science Editors (CSE) style, which is quite different from the MLA or Chicago Style you may be accustomed to using in the humanities. The CSE recommends the following three systems:

- Citation-Sequence
- Name-Year
- Citation-Name

The Citation-Name system is a hybrid of the other two and will be discussed briefly in Chapter 4.

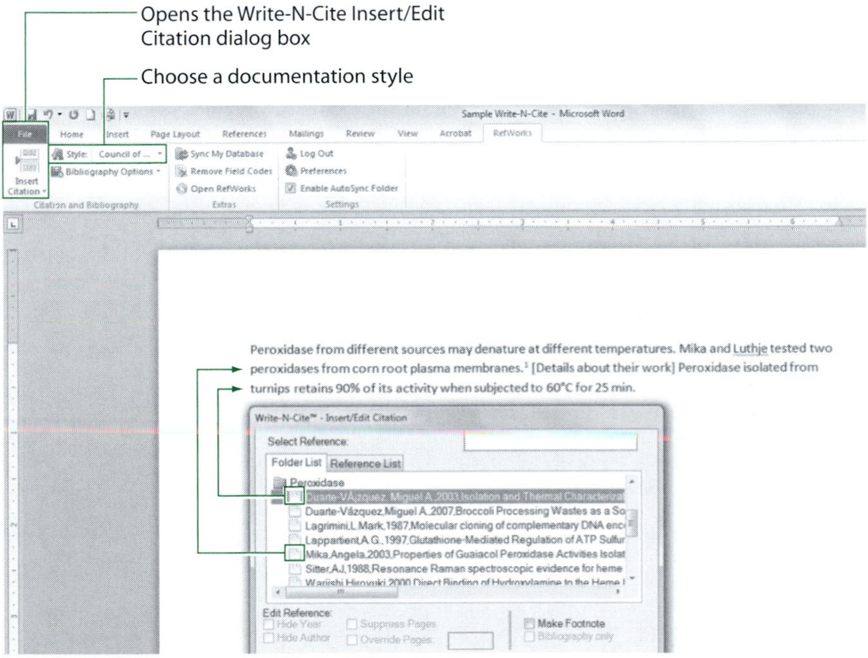

Figure 2.10 To insert a citation in C-S style, position the cursor in the Word document after the period. Click **Insert Citation** and navigate to the folder that contains the reference. After clicking the desired reference and then **OK**, Write-N-Cite inserts a superscripted number.

CITATION-SEQUENCE (C-S)

In the Citation-Sequence system, in-text citations are numbered sequentially and the corresponding full reference is given in a numbered list at the end of the paper.

1. Begin typing your paper in Word. Save the document after you come to a sentence in which you want to cite a reference.

2. Click the **RefWorks** tab and log in.

3. Select the **Council of Science Editors — CSE 7th, Citation-Sequence** style from **RefWorks | Citation & Bibliography | Style | Select Other Style** (Figure 2.10).

4. Position the cursor *after the period* and click **RefWorks | Citation & Bibliography | Insert Citation**. In the **Write-N-Cite Insert/Edit Citation** dialog box, navigate to the relevant folder and click the reference that is to be cited. A superscripted number will appear in the Word document.

5. Repeat this process for each reference to be cited.

6. Save the document just before you are ready to generate the end reference list. This step is important, because Write-N-Cite will not properly format the in-text citation and the end references list if the document has not been saved.

7. Position the cursor at the end of the document. Click **RefWorks | Citation & Bibliography | Bibliography Options | Insert Bibliography** (Figure 2.11).

8. In your Word document, in-text citations are listed sequentially and the information in the end references is in the correct order (see Figure 2.11). Think of the time you'll save by not having to type reference lists!

NAME-YEAR (N-Y)

In the Name-Year system, the in-text citation is given in the form of author and year. The number of authors determines the format of the citation:

- 1 author: Author's last name followed by year of publication
- 2 authors: First author's last name and second author's last name followed by year of publication
- 3 or more authors: First author's last name followed by the words *and others* (or *et al.*) and year of publication

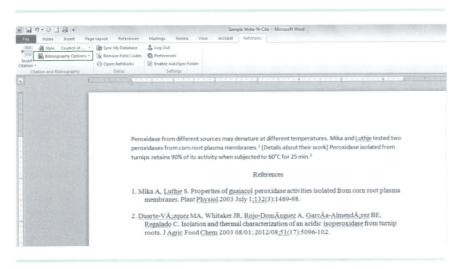

Figure 2.11 Final appearance of a sample lab report formatted using the Citation-Sequence system. After clicking **Bibliography Options | Insert Bibliography**, Write-N-Cite generates the end reference list based on the style selected.

The corresponding full references are listed alphabetically at the end of the paper.

1. Begin typing your paper in Word. Save the document after you come to a sentence in which you want to cite a reference.

2. Click the **RefWorks** tab and log in.

3. Select the **Council of Science Editors - CSE 7th, Name-Year Sequence** style from **RefWorks | Citation & Bibliography | Style | Select Other Style** (see Figure 2.10).

4. Position the cursor *ahead of the period* and click **RefWorks | Citation & Bibliography | Insert Citation**. In the **Write-N-Cite Insert/Edit Citation** dialog box, navigate to the relevant folder and click the reference that is to be cited.

5. Repeat this process for each reference to be cited.

6. Save the document just before you are ready to generate the end reference list. This step is important, because Write-N-Cite will not properly format the in-text citation and the end references list if the document has not been saved.

7. Position the cursor at the end of the document. Click **RefWorks | Citation & Bibliography | Bibliography Options | Insert Bibliography**.

Figure 2.12 Final appearance of a sample lab report formatted using the Name-Year system. After clicking **Bibliography Options | Insert Bibliography**, Write-N-Cite generates the end reference list based on the style selected.

8. In your Word document, in-text citations are listed sequentially and the information in the end references is in the correct order (Figure 2.12). Think of the time you'll save by not having to type reference lists!

Chapter 3

READING AND WRITING SCIENTIFIC PAPERS

No matter whether you are a student or are already engaged in a profession, writing is a fact of life. There are many reasons for writing: to express your feelings, to entertain, to communicate information, and to persuade. When you write scientific papers, your primary reasons for writing are to communicate information and to persuade others of the validity of your methods, findings, and conclusions.

Types of Scientific Writing

Scientific writing takes many forms. As an undergraduate biology major, you will be asked to write laboratory reports, answer essay questions on exams, write summaries of journal articles, and do literature surveys on topics of interest. Upperclass students may write a research proposal for honors work, and then complete their project by submitting an honors thesis. Graduate students typically write master's theses and doctoral dissertations and defend their written work with oral presentations. Professors write lectures, letters of recommendation for students, grant proposals, reviews of articles submitted for publication to scientific journals by their colleagues, and evaluations of grant proposals. In business and industry, scientific writing may take the form of progress reports, product descriptions, operating manuals, and sales and marketing material.

Hallmarks of Scientific Writing

What distinguishes scientific writing from other kinds of writing? One difference is the motive. Scientific writing aims to inform rather than to entertain the reader. The reader is typically a fellow scientist who intends to use this information to

- Stay current in his or her field
- Build on what is already known
- Improve a method or adapt a method to a different research question
- Make a process easier or more efficient
- Improve a product

A second difference is the style. Brevity, a standard format, and proper use of grammar and punctuation are the hallmarks of well-written scientific papers. The authors have something important to communicate, and they want to make sure that others understand the significance of their work. Flowery language and "stream of consciousness" prose are not appropriate in scientific writing because they can obscure the writer's intended meaning.

A third difference between scientific and other types of writing is the tone. Scientific writing is factual and objective. The writer presents information without emotion and without editorializing.

Scientific Paper Format

Scientific papers are descriptions of how the scientific method was used to study a problem. They follow a standard format that allows the reader, first, to determine initial interest in the paper, second, to read a summary of the paper to learn more, and, finally, to read the paper itself for all the details. This format is very convenient, because it allows busy people to scan volumes of information in a relatively short time, then spend more time reading only those papers that truly provide the information they need.

Almost all scientific papers are organized as follows:

- Title
- List of authors
- Abstract
- Introduction
- Materials and Methods
- Results

- Discussion
- Acknowledgments
- References

This standard structure is sometimes called the IMRD format. IMRD is an abbreviation of the core sections of a scientific paper.

The **title** is a **short, informative description of the essence of the paper**. It should contain the fewest number of words that accurately convey the content. Readers use the title to determine their initial interest in the paper.

Only the names of **people who played an active role** in designing the experiment, carrying it out, and analyzing the data appear in the **list of authors**.

The **abstract** is a **summary of the entire paper** in 250 words or less. It contains (1) an introduction (scope and purpose), (2) a short description of the methods, (3) results, and (4) conclusions. There are no literature citations or references to figures in the abstract. If the title sounds promising, readers will use the abstract to determine if they are interested in reading the entire paper.

The **introduction** concisely states what motivated the study, how it fits into the existing body of knowledge, and the objectives of the work. The introduction consists of two primary parts:

1. **Background or historical perspective on the topic**. Primary journal articles and review articles, rather than textbooks and newspaper articles, are cited to provide the reader with direct access to the original work. Inconsistencies, unanswered questions, or new questions that resulted from previous work set the stage for the present study.

2. **Statement of objectives of the work**. What were the goals of the present study?

The **Materials and Methods** section describes, in full sentences and well-developed paragraphs, **how the experiment was done**. The author provides sufficient detail to allow another scientist to repeat the experiment. Volume, mass, concentration, growth conditions, temperature, pH, type of microscopy, statistical analyses, and sampling techniques are critical pieces of information that must be included. When and where the work was carried out is important if the study was done in the field (in nature), but is not included if the study was done in a laboratory. Conventional labware and laboratory techniques that are common knowledge (familiar to the audience) are not explained. In some instances, it is appropriate to use references to describe methods.

The **Results** section is **where the findings of the experiment are summarized**, without giving any explanations as to their significance (the "whys" are reserved for the Discussion section). A good Results section has two components:

- A *text*, which forms the body of this section
- Some form of *visual* that helps the reader comprehend the data and get the message faster than from reading a lengthy description

In the **Discussion** section, the **results are interpreted** and possible explanations are given. The author may:

- Summarize the results in a way that supports the conclusions.
- Describe how the results relate to existing knowledge (literature sources).
- Describe inconsistencies in the data. This is preferable to concealing an anomalous result.
- Discuss possible sources of error.
- Describe future extensions of the current work.

In the **Acknowledgments** section of published research articles, the authors recognize technicians, colleagues, and others who have contributed to the research or production of the paper. In addition, the authors acknowledge the organization(s) that provided funding for the work as well as individuals who provided non-commercially available products or organisms.

References list the **outside sources** the authors consulted in preparing the paper. No one has time to return to a state of zero knowledge and rediscover known mechanisms and relationships. That is why scientists rely so heavily on information published by their colleagues. References are typically cited in the Introduction and Discussion sections of a scientific paper, and the procedures given in the Materials and Methods section are often modifications of those in previous work.

Styles for Documenting References

The Council of Science Editors (CSE Manual 2006) recommends the following three systems for documenting references:

Citation-Sequence System. In the *text*, the source of the cited information is provided in an abbreviated form as a superscripted endnote or a number in square brackets or parentheses. On the *references pages* that follow the Discussion section, the sources are listed in **numerical order** and include the full reference.

Name-Year System. In the *text*, the source is given in the form of author(s) and year. On the *references pages* that follow the Discussion section, the references are listed in **alphabetical order** according to the first author's last name.

Citation-Name System. This system is a hybrid of the Citation-Sequence and Name-Year systems. In the *text*, the source of the cited information is provided in an abbreviated form as a super-scripted endnote or a number in square brackets or parentheses. On the *references pages* that follow the Discussion section, the references are listed in **alphabetical order** according to the first author's last name. The references are then numbered sequentially.

The Name-Year system has the advantage that people working in the field will know the literature and, on seeing the authors' names, will understand the reference without having to check the reference list. This system is more commonly used and generally is preferred. With the Citation-Sequence and Citation-Name systems, for each reference the reader must turn to the reference list at the end of the paper to gain the same information. Both systems are described in detail on pp. 81–94.

Strategies for Reading Journal Articles

Papers in scientific journals are written by experts in the field. Because you are not yet an expert, you will probably find it difficult to read and understand journal articles. The following strategy may help.

Determine the topic. First, try to determine the topic of the article by reading the title, the abstract, and the first few sentences of the introduction. Based on the keywords, what do you expect the paper to be about?

Acquire background information on the topic. Wikipedia is a good place to start, but it should not be considered an authoritative source for academic work. A better choice may be your textbook, written by scientists and reviewed by other scientists before publication. Because textbook authors generally write for a student audience, not a group of experts, your textbook is likely to be easier to read than the primary literature. See "Strategies for Reading your Textbook" on pp. 37–39 for some ways to read biology textbooks efficiently.

Read the introduction. The introduction is usually easier to follow than the abstract. The first few sentences are aimed at attracting reader interest and the topic is introduced in broad terms. Subsequent sentences narrow down the topic and the specific goals of the paper are presented at the end of this section. Skim the introduction with the following questions in mind:

- Why were the authors interested in this topic or problem?
- What was known about the topic?
- What was unknown or what questions were the authors interested in answering?
- Did the authors propose any hypotheses?
- What are the objectives of the current work?

Read the Results section selectively. Look at the figures and tables to determine what variables were studied. The independent variable (the one the investigator manipulated) is plotted on the x-axis, and the dependent variable (the one that changes in response to the independent variable) is plotted on the y-axis. Also look for variables in column headings of tables.

Look for a qualitative description of figures and tables in the figure/table caption and in the body of the Results section (text). The caption states the main idea of the visual. The topic sentence of the paragraph in the text does the same. Supporting sentences follow, providing details on what trends or findings the reader should notice in each visual. When you read about the results, some questions to consider are:

- What kind of data are presented: descriptive or numerical?
- If a hypothesis was tested, was there a difference between the controls and the treatment groups?
- Looking at the graphs, what was the relationship between the independent and dependent variables?
- What is the subject of photos and images?

If necessary, reread the introduction to recall the main objectives of the work. Try to understand the big picture before concerning yourself with the details.

Read the Discussion section. A good Discussion section is structured like a triangle, narrow at the top and wide at the base: the information flows from specific to broad (just the opposite of the introduction). In the first few paragraphs, the author interprets the findings of the current work, tying the results directly to the question or problem posed in the introduction. For example, if the goal articulated in the introduction was to determine if hydroxylamine acts as a competitive inhibitor in the peroxidase–hydrogen peroxide reaction, then the reader expects to find the answer to that question in the Discussion section. Each assertion is backed up with experimental evidence.

In the next part of the discussion, the author compares the results described in his/her paper to those in published articles. If the results do not agree, then the author tries to explain why they do not agree. Finally, the author may discuss the implications of the results for our understanding of broader issues and describe future research.

Skim the Materials and Methods section. Scan the subheadings (if present) and the topic sentence of each paragraph to identify the basic approach. Are you familiar with any of the methods? Try to understand the overall concept and do not concern yourself with the details at this stage.

Expect to read the article several times. Even experts may read journal articles several times before they understand the methodology and the implications of the findings. When reading an article for the first time, your mind is exposed to new concepts. When you stop reading, your unconscious mind has time to process these concepts. Then, when you read the article again, a day or a week or a month later, some things that seemed incomprehensible on the first reading may make more sense.

Be an active reader. Instead of using a highlighter, take notes in your own words (see p. 41). Jot down questions where something is unclear. Forcing yourself to engage actively with the words on the page helps you clarify what you know and what you don't know. Active reading is hard work and requires your full attention. Instead of taking the multitasking approach and trying to read, text, email, and socialize with your friends all at the same time, turn off your phone and email and concentrate on the reading for a full hour. You will be surprised at how much more you can accomplish, especially if you set a time limit and plan to do something pleasurable afterwards.

Strategies for Reading Your Textbook

The expectation to read journal articles more than once also applies to reading chapters in your textbook. Repetition is a key ingredient in learning the material. Repetition not only provides you with multiple opportunities to be exposed to the material, but also gives you time to digest it. The basic approach is to read for organization and key concepts first, and then to fill in the details with each subsequent reading. Remember that each time you read the material, you will learn a little more.

The two strategies described here work best with a chapter or section of text no longer than 25–30 pages. The first strategy is proposed by Counselling Services at the University of Victoria, Canada (Palmer-Stone 2001).

1. Take no more than 25 minutes to:

- Read the chapter title, introduction, and summary (at the end of the chapter, if present)

- Read the headings and subheadings
- Read the chapter title, introduction, summary, headings, and subheadings again
- Skim the topic sentence of each paragraph (usually the first or second sentence)
- Skim italicized or boldfaced words

2. Close your textbook. Take a full 30 minutes to:
   - Write down everything you can remember about what you read in the chapter (make a "mind map"). Each time you come to a dead end, use memory techniques such as associating ideas from your reading to lecture notes or other life experiences; visualizing pages, pictures, or graphs; staring out the window to daydream; and letting your mind go blank.
   - Figure out how all this material is related. Organize it according to what makes sense in your mind, not necessarily according to how it is organized in the textbook. Write down questions and possible contradictions to check on later.

3. Open your textbook. Fill in the blanks in your mind map with a different colored pencil.

4. Read the chapter again, this time normally. Make another mind map.

A second strategy is:

1. Skim the chapter title, headings, and subheadings for an overview of the chapter content. Write down the headings and subheadings in the form of an outline.

2. Look at your outline and ask yourself the following questions:
   - What is the main topic of this chapter?
   - How do each of the headings relate to the topic?
   - How does each subheading relate to its heading?

3. Read each section, paying special attention to the topic sentence of each paragraph. At the end of each section, summarize the content in your own words. Answer the following questions:
   - What's the point?
   - What do I understand?
   - What is confusing?

4. If you read the assigned pages before the lecture, you can pay attention to the lecture content instead of just frantically taking notes. Your instructor may provide PowerPoint slides for the lecture or your textbook may come with a printed lecture note-book or a DVD with the figures. These aids allow you to spend more time listening and less time writing.

5. After the lecture, while the information is still fresh in your mind, reread your notes on your reading. Ask yourself:

   - What topics did the instructor emphasize in lecture? Fill in your lecture notes with details from your textbook.
   - What material do I understand better now?
   - What questions remain?

Study Groups

If you have read the material several times, taken notes, and listened attentively in lecture, but still have questions, talk about the material with your classmates. Small study groups are one reason why students who choose to major in the sciences persist in the sciences, rather than switching to a non-science major (Light 2001).

What are some benefits of participating in small study groups? One benefit is the comfort level. You may be more likely to talk about problems when you are among your peers; after all, they are not the ones who assign your grade. Secondly, when a group is composed of peers with a similar knowledge base, group members speak the same language. Your instructor speaks a different language, because he or she has already struggled to master the material. When you communicate with your classmates, you verbalize your ideas at a level that is appropriate for your audience of peers. Finally, collaborative learning reflects the way scientists exchange information and share findings in the real world. A spirit of camaraderie develops when people work together toward a common goal. The prospect of learning difficult subject matter is no longer so daunting when you have support from a small group of like-minded individuals. The hard work may even be fun when there is good group chemistry.

Group study is not a substitute for studying alone, however. You must hold yourself accountable for reading the material, taking notes, and figuring out what you do not understand before you meet with your group. If you have not struggled to understand the material yourself, you are not in a position to help a classmate.

Plagiarism

Plagiarism is using someone else's ideas or work without acknowledging the source. Plagiarism is ethically wrong and demonstrates a lack of respect for members of your academic community (faculty and fellow students) and the scientific community in general. Many instructors are now using plagiarism checking services such as Turnitin® and SafeAssign™ by Blackboard to discourage *intentional* plagiarism, such as "borrowing" portions of another student's work, recycling lab reports from previous years, and buying papers on the Internet. Plagiarists who are caught can expect to receive at a minimum a failing grade on the assignment and close scrutiny in subsequent work. Plagiarism may also be cause for expulsion from school.

Many cases of plagiarism are *unintentional*, however, and stem from issues such as

- Failure to understand what kind of information must be acknowledged
- Failure to reference the original material properly
- Failure to understand the subject matter clearly

Information that does not have to be acknowledged

General information that is obtained from sources such as news media, textbooks, and encyclopedias does not have to be acknowledged.

> EXAMPLE: Most of the ATP in eukaryotic cells is produced in the mitochondria.

Information that is common knowledge for your audience does not have to be acknowledged. In an introductory course in cell and molecular biology, for example, students would be expected to know that ATP synthase is the enzyme that produces ATP through oxidative phosphorylation.

> EXAMPLE: ATP is synthesized when protons flow down their electrochemical gradient through a channel in ATP synthase.

Information that has to be acknowledged

Information that falls into any of the following categories must be acknowledged:

- Information that is not widely known
- Controversial statements, opinions, or other people's conclusions

- Pictures or illustrations that you use but did not produce
- Statistics or formulas used in someone else's work
- Direct quotations

Paraphrasing the source text

Direct quotations are used in the humanities, but not in scientific papers. This idiosyncrasy of technical writing requires you to paraphrase the information in the source document. Paraphrasing—using your own words to express someone else's ideas—requires considerable thought and effort on your part. Not only do you have to have sufficient knowledge about the subject, you have to feel comfortable using the vocabulary. Read your textbook and other secondary sources, discuss the topic in your study group, or ask your instructor for clarification. A lot of groundwork has to be done before you can even begin to read a journal article, let alone paraphrase information it contains.

When you have acquired this background information, you are ready to tackle the content. Accept the fact that comprehension is an ongoing process in which you will read the source text, process the information you've read, read the text again, and process some more. When you are comfortable with the content, take notes on the important points, following the collective advice of Hofmann (2010), Lannon and Gurak (2011), McMillan (2012), Pechenik (2012), and other authorities on scientific writing:

- Don't take notes until you have read the source text at least twice and are fairly confident that you have sufficient background on the subject.
- Retain key words.
- Don't use full sentences.
- Use your own words and write in your own style.
- Distinguish your own ideas and questions from those of the source text (e.g., "Me: Applies only to prokaryotes?").
- Use quotation marks to indicate exact or similar wording. Keep in mind that you will have to put the information into your own words if you use the information in your paper.
- Don't cite out of context. Preserve the author's original meaning.
- Fully document the source for later listing in the end references.

Faulty note-taking practices, particularly those that involve copying large portions of the original text, are likely to result in unintentional plagiarism. Beware of the pitfalls illustrated in Table 3.1. To avoid plagiarism, write in your own words and cite the source. For practice in identifying and avoiding plagiarism, take Frick's (2004) excellent online plagiarism tutorial. Read your institution's policies on academic responsibility, consult with professionals at your school's writing center, and ask your instructor for help when in doubt.

The Benefits of Learning to Write Scientific Papers

Why is it valuable to learn how to write scientific papers? First, scientific writing is a systematic approach to describing a problem. By writing what you know (and what you do not know) about the problem, it is often possible to identify gaps in your own knowledge.

Second, the scientific method is a logical approach to answering questions. It involves coming up with a tentative solution, gathering information to become more knowledgeable about the topic, evaluating the reliability of the information, testing and analyzing the data, and arriving at a reasonable conclusion. This approach can be applied to many situations in your life, from deciding to which graduate school to apply to choosing your next cell phone or another consumer product.

Third, when you learn to write lab reports, you are investing in your future. Publications in the sciences are affirmation from your colleagues that your work has merit; you have been accepted into the community of experts in your field. Even if your career path is not in the sciences, scientific writing is very logical and organized, characteristics appreciated by busy people everywhere.

Credibility and Reputation

The credibility and reputation of scientists are established primarily by their ability to communicate effectively through their written reports. Poorly written papers, regardless of the importance of the content, may not get published if the reviewers do not understand what the writer intended to say.

You should think about your reputation even as a student. When you write your laboratory reports in an accepted, concise, and accurate manner, your instructor knows that you are serious about your work. Your instructor appreciates not only the time and effort required to understand the subject matter, but also your willingness to write according to the standards of the profession.

Model Papers

Before writing your first laboratory report, go to the library and take a look at some biology journals such as *American Journal of Botany, Ecology, The*

TABLE 3.1   Examples of plagiarism

Original Text

F_1 extends from the membrane, with the α and β subunits alternating around a central subunit γ. ATP synthesis occurs alternately in different β subunits, the cooperative tight binding of ADP + P_i at one catalytic site being coupled to ATP release at a second. The differences in binding affinities appear to be caused by rotation of the γ subunit in the center of the $\alpha3\ \beta3$ hexamer.

Plagiarized Text	Reason
According to Fillingame (1997), F1 extends from the membrane, with the α and β subunits alternating around a central subunit γ. ATP synthesis occurs alternately in different β subunits, the cooperative tight binding of ADP + P_i at one catalytic site being coupled to ATP release at a second. The differences in binding affinities appear to be caused by rotation of the γ subunit in the center of the $\alpha3\ \beta3$ hexamer.	The author's actual words were used without quotation marks or indenting the citation. Because direct quotations are not used in scientific papers, it is imperative that you paraphrase. Using the original text is plagiarism even when the source is cited.
F_1 consists of α and β subunits alternating around a central subunit γ. In the β subunits, tight binding of ADP + P_i occurs at one catalytic site and ATP is released at a second. The different binding affinities may be caused by rotation of the γ subunit in the center (Fillingame 1997).	The basic sentence structure of the original text was maintained. A few words were omitted or changed, but the text is still highly similar to the original.
ATP synthase consists of a transmembrane protein (F_o), a central shaft (γ), and an F_1 head made up of α and β subunits. As protons enter F_u, the shaft rotates, changing the conformation of the β subunits, allowing ADP and P_i to bind and be released as ATP.	The text was paraphrased, but the source of the information was not cited.

Source: From Fillingame RH. 1997. Coupling H^+ transport and ATP synthesis in F_1F_o-ATP synthases: glimpses of interacting parts in a dynamic molecular machine. *The Journal of Experimental Biology* [Internet] [cited 2012 Oct 30]; 200: 217–224. Available from: http://jeb.biologists.org/content/200/2/217.full.pdf+html

EMBO Journal, Journal of Biological Chemistry, Journal of Molecular Biology, and *Marine Biology.* Photocopy one or two journal articles that interest you so you can refer to them for format questions.

Almost all journals devote one page or more to "Instructions to Authors," in which specific information is conveyed regarding length of the manuscript, general format, figures, conventions, references, and so on. Skim this section to get an idea of what journal editors expect from scientists who wish to have their work published.

Because most beginning biology students find journal articles hard to read, a sample student laboratory report is given in Chapter 6. Read the comments in the margins as you peruse the report to familiarize yourself with the basics of scientific paper format and content, as well as purpose, audience, and tone.

Chapter 4

Step-by-Step Instructions for Preparing a Laboratory Report or Scientific Paper

In order to prepare a well-written laboratory report according to accepted conventions, the following skills are required:

- A solid command of the English language
- An understanding of the scientific method
- An understanding of scientific concepts and terminology
- Advanced word processing skills
- Knowledge of computer graphing software
- The ability to read and evaluate journal articles
- The ability to search the primary literature efficiently
- The ability to evaluate the reliability of Internet sources

If you are a first- or second-year college student, it is unlikely that you possess all of these skills when you are asked to write your first laboratory report. Don't worry. The instructions in this chapter will guide you through the steps involved in preparing the first draft of a laboratory report. Revision is addressed in the next chapter, and the Appendices will help you with word processing and graphing tasks.

Timetable

Preparing a laboratory report or scientific paper is hard work. It will take much more time than you expect. Writing the first draft is only the first step. You must also allow time for editing and proofreading (revision). If you work on your paper in stages, the final product will be much better than if you try to do everything at the last minute.

TABLE 4.1	Timetable for writing your laboratory report	
Time Frame	**Activity**	**Rationale**
Day 1	Complete laboratory exercise.	It's fun. Besides, you need data to write about.
Days 2–3	Write first draft of laboratory report.	The lab is still fresh in your mind. You also need time to complete the subsequent tasks before the due date.
Day 4	Proofread and revise first draft (hard copy).	Always take a break after writing the first draft and before revising it. This "distance" gives you objectivity to read your paper critically.
Day 5	Give first draft to a classmate for review.	Your peer reviewer is a sounding board for your writing. He/she will give you feedback on whether what you intended to write actually comes across to the reader. You may wish to alert your peer reviewer to concerns you have about your paper (see "Get Feedback" in Chapter 5).
	Arrange to meet with your classmate after he/she has had time to review your paper ("writing conference").	An informal discussion is useful for providing immediate exchange of ideas and concerns.

The timetable outlined in Table 4.1 breaks the writing process down into stages, based on a one-week time frame. You can adjust the time frame according to your own deadlines.

Format your report correctly

Although content is important, the appearance of your paper is what makes the first impression on the reader. If the pages are out of order and the ink is faded, subconsciously or not, the reader/evaluator is going to associate a sloppy paper with sloppy science. You cannot afford that kind of reputation. In order for your work to be taken seriously, your paper has to have a professional appearance.

Scientific journals specify the format in their "Instructions to Authors" section. If your instructor has not given you specific instructions, the layout specified in Table 4.2 will give your paper a professional look.

TABLE 4.1	*Continued*	
Time Frame	Activity	Rationale
Day 6	Peer reviewer reviews laboratory report.	The peer reviewer should review the paper according to two sets of criteria. One is the conventions of scientific writing as described in "Scientific Paper Format" in Chapter 3, and the other is the set of questions in "Get Feedback" in Chapter 5.
	Hold writing conference during which the reviewer returns the first draft to the writer.	An informal discussion between the writer and the reviewer is useful to give the writer an opportunity to explain what he/she intended to accomplish, and for the reviewer to provide feedback.
Days 6–7	Revise laboratory report.	Based on your discussion with your reviewer, revise as necessary. Remember that you do not have to accept all of the reviewer's suggestions.
Day 8	Hand in both first draft and revised draft to instructor.	Your instructor wants to know what you've learned (we never stop learning either!).

Consult the sample "good" student laboratory report in Chapter 6 for an overview of the style and layout. An electronic file called "Biology Lab Report Template," available at <http://sites.sinauer.com/Knisely4e> is formatted according to the guidelines of Table 4.2 and provides prompts that help you get started writing in scientific paper format. For details on how to format documents in Microsoft Word, see the "Commands in Word 2010" section in Appendix 1.

Computer savvy

Know your PC and your word processing software. Most of the tasks you will encounter in writing your laboratory report are described in Appendix 1, "Word Processing in Microsoft Word 2010" and Appendix 2, "Making Graphs in Microsoft Excel 2010 and Excel for Mac 2011." If there is a task that is not covered in these appendices, write it down and ask an expert later.

TABLE 4.2	Instructions to authors of laboratory reports
Feature	**Layout**
Paper	8½" x 11" (or DIN A4) white bond
Margins	1.25" left and right; 1" top and bottom
Font size	12 pt (points to the inch)
Typeface	Times Roman or another *serif* font. A serif is a small stroke that embellishes the character at the top and bottom. The serifs create a strong horizontal emphasis, which helps the eye scan lines of text more easily.
Symbols	Use word processing software. Do not write symbols in by hand.
Pagination	Arabic number, top right on each page except the first
Justification	Align left/ragged right or Full/even edges
Spacing	Double
New paragraph	Indent 0.5"
Title page (optional)	Title, authors (your name first, lab partner second), class, and date
Headings	Align headings for Abstract, Introduction, Materials and Methods, Results, Discussion, and References on left margin or center them. Use consistent format for capitalization. Do not start each section on a new page unless it works out that way coincidentally. Keep section heading and body together on the same page.
Subheadings	Use sparingly and maintain consistent format.
Tables and figures	Incorporate into text as close as possible after the paragraph where they are first mentioned. Use descriptive titles, sequential numbering, proper position above or below visual. May be attached on separate pages at end of document, but must still have proper caption. Keep table/figure and its caption together on the same page.
Sketches	Hand-drawn in pencil or ink. Other specifications as in "Tables and figures" above.
References	Citation-Sequence System: Make a numbered list in order of citation.
	Name-Year System: List references in alphabetical order by the first author's last name. Use a hanging indent (all lines but the first indented) to separate individual references.
	Both systems: Use accepted punctuation and format.
Assembly	Place pages in order, staple top left.

If you run into a major problem that prevents you from using your PC, you should have a backup plan in place (familiarity with another PC).

Always back up your files somewhere other than your computer's hard drive. Options may include a USB flash drive (also called a jump drive or thumb drive), an external hard drive, or online. Online options include cloud services such as Google Drive, saving your files to your organization's server, or emailing files to yourself. See the section "Backing up your files" on pp. 198–200 for more information.

Save your file frequently while writing your paper by clicking 🖫 on the Quick Access Toolbar. You can also adjust the settings for automatically saving your file by clicking **File | Options | Save | Save AutoRecover information every __ minutes**.

Install antivirus software on your computer and always check flash drives for viruses before you use them. Beware of files attached to email messages. Do not open attachments unless you are sure they come from a reliable source.

Store flash drives with their caps on to keep dust out. Protect them from excess humidity, heat, and cold. Only remove a flash drive from a computer after you eject it and the message "Safe to Remove Hardware" is displayed.

If you must eat and drink near a computer, keep beverages and crumbs away from the hard drive and keyboard.

Getting Started

Set aside 1 hour to begin writing the laboratory report as soon as possible after doing the laboratory exercise. Turn off your cell phone and get off Facebook, Twitter, YouTube, and email. Writing lab reports requires your full concentration. What matters is the quality, not the quantity, of time you spend on your assignments. Promise yourself a reward for time well spent.

Reread the laboratory exercise

You cannot begin to write a paper without a sense of purpose. What were the objectives of your experiment or study? What questions are you supposed to answer? Take notes on the laboratory exercise to prevent problems with plagiarism when you write your laboratory report.

Organization

If your instructor provided a rubric or other instructions for organizing your lab report, follow the instructions exactly. Otherwise use the standard IMRD format, as described in Chapter 3.

Audience

Scientific papers are written for scientists. Similarly, laboratory reports should be written for an audience of fellow student-biologists, who have a knowledge base similar to your own. When deciding how much background information to include, assume that your audience knows only what you learned in class. Use scientific terminology, but define any terms known only to experts ("jargon").

Write for an audience of fellow scientists, not students in a classroom situation. Note the difference between the original text and the revision in the following examples:

FAULTY: The experiments performed by the students dealt with how different wavelengths of light affect seed germination.

REVISION: The purpose of the experiment was to determine how different wavelengths of light affect seed germination.

FAULTY: The purpose of this experiment is to become acquainted with new lab techniques such as protein analysis, serial dilutions, and use of the spectrophotometer.

REVISION: The purpose of this experiment was to use the biuret assay to determine protein concentration in egg white.

Writing style

Laboratory reports are formal written assignments. Avoid slang and contractions and choose words that reflect the serious nature of scientific study. Readers of scientific papers trust the scientific method and are confident that the facts speak for themselves. For this reason, write objectively—that is, do not make judgments. When making a statement that may not be obvious to the audience, always back it up by citing an authoritative source or by providing experimental evidence. Because the focus is on the science, not the scientist, passive voice is used more frequently (especially in the Materials and Methods section) than in other kinds of writing. Use active voice in the other sections, however, because it makes sentences shorter and more dynamic.

Past and present tense have specific connotations in scientific papers. Authors use present tense to make *general statements* that the scientific community agrees are valid. Statements that are generally valid include explanations of phenomena based on experimental results that have been replicated by many scientists. Therefore, use present tense in the Introduction and Discussion sections when describing information accepted by the scientific community, and cite the source of any information that is not common knowledge for your audience. On the other hand, authors use past tense to limit interpretations and conclusions to *their own work*. For this reason, use past tense in the Materials and Methods and Results sections, and whenever you are describing work that you personally carried out.

Start with the Materials and Methods Section

The order in which you write the different sections is not the order in which they appear in the finished laboratory report. The rationale for this plan will become obvious as you read on. The Materials and Methods section requires the least amount of thought, because you are primarily restating the procedure in your own words.

Tense

When you write your laboratory report, describe the procedure in *past*, not present, tense because (1) these are completed actions and (2) you are describing your own work. Do *not* copy the format of your laboratory exercise, in which the instructions may be arranged in a numbered list and the imperative (command) form of verbs may be used for clarity.

Voice

There are two grammatical voices in writing: active and passive. In active voice, the subject *performs* the action. In passive voice, the subject *receives* the action. Passive voice is preferred in the Materials and Methods section because the subject that receives the action is more important than who performed it. The logic is that anyone with the appropriate training should be able to perform the action. Consider the following examples:

ACTIVE VOICE: I peeled and homogenized the potatoes.

PASSIVE VOICE: The potatoes were peeled and homogenized.

The sentence written in active voice is more natural and dynamic, but it shifts the emphasis from the subject, "the potatoes" to "I." Passive voice places the emphasis on the potatoes, where it belongs. Because sentences written in passive voice tend to be longer and less direct than those written in active voice, try to use active voice when the performer (you) is not the subject of the sentence.

Level of detail

A well-written Materials and Methods section will *provide enough detail to allow someone with appropriate training to repeat the procedure*. For example, for a **molecular biology** procedure, include essential details such as the concentration and pH of solutions, reaction and incubation times, volume, temperature, wavelength (set on a spectrophotometer), centrifugation speed, dependent and independent variables, and control and treatment groups. On the other hand, *do not describe routine lab procedures* such as:

- How to calculate molarity or use $C_1V_1 = C_2V_2$ to make solutions
- Taring a balance before use
- Using a vortex mixer to ensure that solutions are well mixed
- Describing how to zero (blank) a spectrophotometer before measuring the absorbance of the samples
- Explaining what type of serological pipette or micropipettor is appropriate for a particular volume
- Designating the type of flasks or beakers to use
- Specifying the duration of the entire study ("In our two-week experiment, …")

For a **field experiment**, however, time *is* important. When observing or collecting plants and animals in nature, be sure to include in the Materials and Methods section time of day, month, and year; sampling frequency; location and dimensions of the study site; sample size; and statistical analyses. Depending on the focus of your lab report, it may also be prudent to describe the geology, vegetation, climate, natural history, and other characteristics of the study site which could influence the results.

Here are some guidelines for the level of detail to include in the Materials and Methods section.

Not enough information. Include all relevant information needed to repeat the experiment.

FAULTY: In this lab, we mixed varying amounts of BSA stock solution with varying amounts of TBS using a vortex mixer. We used a spectrophotometer to measure absorbance of the 4 BSA samples, and then we determined the concentration of 4 dilutions of egg white from the standard curve.

EXPLANATION: This procedure does not give the reader enough information to repeat the experiment, because essential details like *what concentrations of BSA* were used to construct the standard curve, *what dilutions of egg white* were tested, and the *wavelength* set on the spectrophotometer have been left out.

REVISION: Bovine serum albumin (BSA) solutions (2, 3, 5, 10 mg/mL) were prepared in tris-buffered saline (TBS). The egg white sample was serially diluted 1/5, 1/15, 1/60, and 1/300 with TBS. The absorbance of all samples was measured at 550 nm using a Spec 20 spectrophotometer.

The following are examples of **too much information**.

Do not list materials and methods separately. The wording of the section heading makes it tempting to separate the content into two parts. In fact, **materials should not be listed separately** unless the strain of bacteria, vector (plasmid), growth media, or chemicals were obtained from a special or noncommercial source. It will be obvious to the reader what materials are required on reading the methods.

Describe the solutions, not the containers.

FAULTY: Eight clean beakers were labeled with the following concentrations of hydrogen peroxide and those solutions were created and placed in the appropriate beaker: 0, 0.1, 0.2, 0.5, 0.8, 1.0, 5.0, and 10.0.

EXPLANATION: Using clean, suitable containers to store solutions is common practice in the laboratory. Putting labels on labware is also a routine procedure. An essential detail missing from this sentence is the units.

REVISION: The following hydrogen peroxide solutions were prepared: 0, 0.1, 0.2, 0.4, 0.8, 1.0, 5.0, and 10.0%.

Specify the concentrations, not the procedure for making solutions.

FAULTY: To make the dilution, a micropipette was used to release 45, 90, 135, and 180 µL of bovine serum albumin (BSA) into four different test tubes. To complete the dilution, 255, 210, 165, and 120 µL of TBS was added, respectively.

EXPLANATION: With appropriate instruction, making dilutions of stock solutions becomes a routine procedure. In the above example, you should assume that your readers can make the solution using the appropriate measuring instruments as *long as you specify the final concentration.*

REVISION: The following concentrations of BSA were prepared for the Bradford assay: 300, 600, 900, and 1200 µL/mL.

Include only essential procedures and write concisely.

FAULTY: The test tubes were carried over to the spectrophotometer and the wavelength was set to 595 nm (nanometer). The spectrophotometer was zeroed using the blank. Each of the remaining 8 samples in the test tubes were individually placed into the empty spec tube, which was then placed in the spectrophotometer where the absorbance was determined.

EXPLANATION: The only detail important enough to mention is the wavelength.

REVISION: The absorbance of each sample was measured with a Spectronic 20 spectrophotometer at 595 nm.

Avoid giving "previews" of your data analysis.

FAULTY: A graph was plotted with Absorbance on the *y*-axis and Protein concentration on the *x*-axis. An equation was found to fit the line, then the

unknown protein absorbances that fell on the
graph were plugged into the equation, and a
concentration was found.

EXPLANATION: Making graphs is something that you do when
you analyze your raw data, but it is not part of
the experimental procedure. How and why you
chose to organize the data will become obvious
to the reader in the Results section, where you
display graphs, tables, and other visuals and
describe the noteworthy findings.

REVISION: Delete this entire passage.

Cite published sources. If you are paraphrasing a published laboratory exercise, it is necessary to cite the source (see "Documenting Sources" on pp.
81–94). Unpublished laboratory exercises are not usually cited; ask your
instructor to be sure.

Do the Results Section Next

The Results section is a *summary* of the key findings of your experiment.
This section has two components:

- Visuals, such as tables and figures
- A body, or text, in which you describe the results shown in the
  visuals

When you work on the Results section, you will complete the following
tasks, which are often done concurrently, not necessarily sequentially:

- **Analyze the raw data.** Raw data are all the observations and
  measurements that you recorded in your lab notebook. It is
  your job as the author to analyze all these data and process the
  information for the reader. **Do not simply transfer raw data
  into your lab report** (your instructor may ask you to attach
  pages from your lab notebook as an appendix, however).
  Instead, summarize the data by eliminating aberrant results
  (because you realized that you made a mistake in obtaining
  these results), averaging replicates, using statistical methods to
  see possible trends, and/or selecting representative pictures
  (micrographs or gel images). The goal of data analysis in gen-

eral is to try to figure out what the data show. More specifically, you compare the results to the predictions that were based
on the hypotheses you proposed when designing your experiment. When the results match the predictions, then the
hypotheses are supported. Conversely, when the results are
unexpected, further research may be required.

- ■ **Organize summarized data in tables or figures.** When you
  organize summarized data in a table or plot numerical values
  on a graph, you may be able to see trends that were not apparent before. Effective visuals are more powerful than words
  alone and they provide strong support for your arguments. See
  the "Organizing Your Data" section.

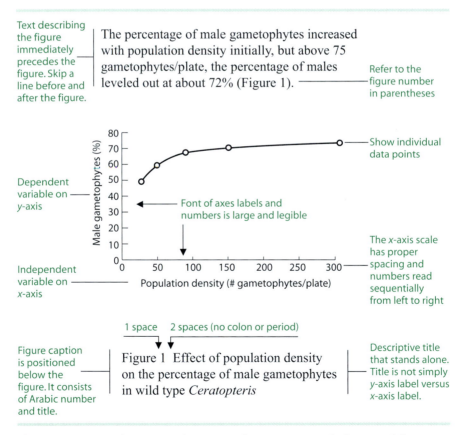

Text describing the figure immediately precedes the figure. Skip a line before and after the figure.

The percentage of male gametophytes increased with population density initially, but above 75 gametophytes/plate, the percentage of males leveled out at about 72% (Figure 1).

Refer to the figure number in parentheses

Show individual data points

Dependent variable on y-axis

Font of axes labels and numbers is large and legible

Independent variable on x-axis

The x-axis scale has proper spacing and numbers read sequentially from left to right

1 space   2 spaces (no colon or period)

Figure caption is positioned below the figure. It consists of Arabic number and title.

Figure 1 Effect of population density on the percentage of male gametophytes in wild type *Ceratopteris*

Descriptive title that stands alone. Title is not simply y-axis label versus x-axis label.

Figure 4.1 Excerpt from a Results section showing a properly formatted figure
with one line (data set); text that describes the figure precedes it.

- **Decide in which order to present the tables and figures.** The sequence should be logical, so that the first visual provides a basis for the next or so that the reader can easily follow your line of reasoning.
- **Describe each visual in turn and refer to it in parentheses.** Describe the most important thing you want the reader to notice about the visual. Refer to the visual by number in parentheses at the end of the first sentence in which you describe it. Position the visual after the descriptive paragraph (Figure 4.1). Other examples of good descriptions of visuals are given in Figures 4.2 and 4.4.

Writing the body of the Results section

The body of the Results section is a description of the important findings of your study or experiment.

Tense. Use *past* tense, since you are referring specifically to your own results. Do not use present tense, because present tense would indicate that the scientific community has already agreed that your results are generally valid.

Voice. Use active voice as much as possible to make your descriptions clear and concise.

Style. Refer to Figure 4.1 as you read the following examples, which illustrate some of the common problems you may encounter when writing your Results section.

Report the results objectively, **without explaining or interpreting** them.

FAULTY: As population density increased, the percentage of male gametophytes also increased due to the effect of the pheromone antheridiogen.

EXPLANATION: Do not try to explain the results in the Results section. Save interpretation for the discussion.

REVISION: The percentage of male gametophytes increased with population density initially, but above 75 gametophytes/plate, the percentage of males leveled out at about 72% (Figure 1).

Make every sentence meaningful.

FAULTY: The results of the testing showed that population density affected the percentage of male gameto-phytes.

EXPLANATION: *How* did population density affect the percentage of males?

FAULTY: The final analyzed data from the lab showed that there was a population density at which the number of gametophytes was no longer increas-ing.

EXPLANATION: *What was the density* at which the percentage of males no longer increased?

FAULTY: Figure 1 represents the effect of population den-sity on the percentage of male gametophytes. The class data was averaged for the effect that each density had on the number of males.

EXPLANATION: *What was the effect* of population density on the percentage of males?

REVISION: The percentage of male gametophytes increased with population density initially, but above 75 gametophytes/plate, the percentage of males leveled out at about 72% (Figure 1).

Similarly, **eliminate unnecessary introductory phrases** such as:

- It was found that…
- The results show that…
- There is a general trend where…
- It can be determined that…
- When the test plates were observed and counted, …
- When the values for each calculation had been obtained, it was clear that…

REVISION: Delete the introductory phrase and begin the sentence with an actual result.

Always **refer to the visual** that contains the data that you are describing.

FAULTY: The results that were obtained while completing the lab clearly showed that the percentage of male gametophytes increased with population density initially, but above 75 gametophytes/plate, the percentage of males leveled out at about 72%.

FAULTY: The figure below shows that there was a linear relationship between population density and percentage of male gametophytes up to about 75 gametophytes/plate. At higher densities, the percentage of males stayed the same.

EXPLANATION: *Which figure* contains the results the author is describing?

REVISION: The percentage of male gametophytes increased with population density initially, but above 75 gametophytes/plate, the percentage of males leveled out at about 72% (Figure 1).

Equations

Equations are technically part of the text and should *not* be referred to as figures. Equations are set off from the rest of the text on a separate line. If you have several equations and need to refer to them unambiguously in the body of the Results (or another) section, number each equation sequentially and place the number in parentheses on the right margin. For example:

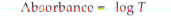

$$\text{Absorbance} = \log T \tag{1}$$

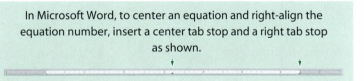

In Microsoft Word, to center an equation and right-align the equation number, insert a center tab stop and a right tab stop as shown.

If you are presenting a sequence of calculations, align the = symbol in each line, as in the following example:

Protein concentration of the unknown sample was determined using the equation of the biuret standard curve. The measured

Graph	Purpose	Example
Histogram	To show the distribution of a quantitative variable.	Distribution of grades on an exam. Y-axis shows number of students; x-axis shows numerical score on the exam.
Scatterplot	To show the relationship between two quantitative variables measured on the same individuals. Look for an overall pattern and for deviations from that pattern. If the points lie close to a straight line, a linear trendline may be superimposed on the scatter graph. The correlation, r, indicates the strength of the linear relationship.	Relationship between shell length and mass. If we are just looking for a pattern, it doesn't matter which variable is plotted on which axis. If we suspect that mass depends on length, plot mass on the y-axis and length on the x-axis. Look at the form, direction, and strength of the relationship.
Line graph	To show the relationship between two quantitative variables. One variable may be dependent on the other. The variable that is being manipulated is called the independent or explanatory variable. The variable that changes in response to the independent variable is called the dependent or response variable. By convention, the independent variable is plotted on the x-axis and the dependent variable is plotted on the y-axis. Error bars may be included to show variability.	Relationship between enzyme activity and temperature. Because temperature is the variable that is being manipulated, it is plotted on the x-axis. Because enzyme activity is the response being measured, it is plotted on the y-axis.

TABLE 4.3   Types of graphs and their purpose

absorbance value was substituted for y, and the equation was solved for x (the protein concentration):

$$y = 0.0417x$$
$$0.225 = 0.0417x$$
$$5.40 = x$$

Thus, the protein concentration of the sample was 5.40 mg/mL.

TABLE 4.3 *Continued*		
Graph	**Purpose**	**Example**
Scatterplot with regression line	To predict the value of y for a given value of x or vice versa. The response variable must be dependent on the explanatory variable and the relationship must be linear. The regression line takes the form $y = mx + b$, where m is the slope and b is the y-intercept.	Standard curve for a protein assay. Protein concentrations of a standard such as BSA are plotted on the x-axis. Absorbance (measured by a spectrophotometer) for each concentration is plotted on the y-axis. A regression line is fitted to the data. To predict the protein concentration of a sample (x), measure its absorbance (y) and solve the regression equation for x.
Bar graph	To show the distribution of a categorical (non-quantitative) variable.	Effect of different treatments on plant height. One axis shows the treatment category and the other shows the numerical response.
Pie graph	To show the distribution of a categorical (non-quantitative) variable in relation to the whole. All categories must be accounted for so that the pie wedges total 100%.	Composition of insects in a backyard survey. Each wedge represents the percentage of an order of insects. Orders with low representation may be combined into an "Other" wedge to complete the pie.

In Microsoft Word 2008 or later, type equations using the Equation Tool, accessed by clicking **Insert | Symbols | Equation**. Type the first equation into the box. Press **Enter**. Repeat the process for each equation in the group. To align the group of equations on the equal sign, select all of the equations, right-click, and select **Align at =**. This method does not work if each line has a right-aligned equation number. In that case, the equations have to be aligned manually.

Preparing visuals

The most common visuals in scientific writing are tables and figures. A **table** is defined by Webster's dictionary as "a systematic arrangement of data usually in rows and columns for ready reference." A **figure** is any visual that is not a table. Thus, line graphs, bar graphs, pie graphs (also called pie charts), drawings, gel photos, X-ray images, and microscope images are all called *figures* in scientific papers.

The type of visual you use depends on the objectives of your study or experiment and the nature of the data. Use a **table** when

- The exact numbers are more important than the trend.
- Statistics such as sample size, standard error, and P-values are used to support your conclusions.
- Arranging categorical variables and other non-quantitative information makes it easier to interpret the results.

Use a **graph** to show relationships between or among variables. The type of graph that can be used is often dictated by the nature of the variables—quantitative or categorical. **Categorical variables** are groups or categories that have no units of measurement (treatment groups, age groups, habitat, etc.). Bar graphs and pie graphs are commonly used to display results involving categorical variables. **Quantitative variables**, on the other hand, have numerical values with units. Line graphs (also called XY graphs) and scatter graphs (also called scatterplots) display relationships between quantitative variables. Some of the graphs frequently encountered in the field of biology are summarized in Table 4.3 and described individually in the following sections.

Do not feel that you have to have visuals in your lab report. If you can state the results in a sentence, then no visual is needed (see Example 1 in the "Organizing Your Data" section).

Tables

Tables are used to display large quantities of numbers and other information that would be tedious to read in prose. Arrange the categories vertically, rather than horizontally, as this arrangement is easier for the reader to follow (see, for example, Table 1 in Figure 4.2). List the items in a logical order (e.g., sequential, alphabetical, or increasing or decreasing value). Include the units in each column heading to save yourself the trouble of writing the units after each number entry in the table.

By convention, tables in scientific papers do not have vertical lines to separate the columns, and horizontal lines are used only to separate the table caption from the column headings, the headings from the data, and the data from any footnotes. The tables in this book are formatted in this style.

Give each table a caption that includes a number and a title. Center the caption or align it on the left margin *above* the table. Use Arabic numbers, and number the tables consecutively in the order they are discussed in the text. Notice that in this book, the table and figure numbers are preceded by the chapter number. This system helps orient the reader in long manuscripts, but is not necessary in short papers like your laboratory report.

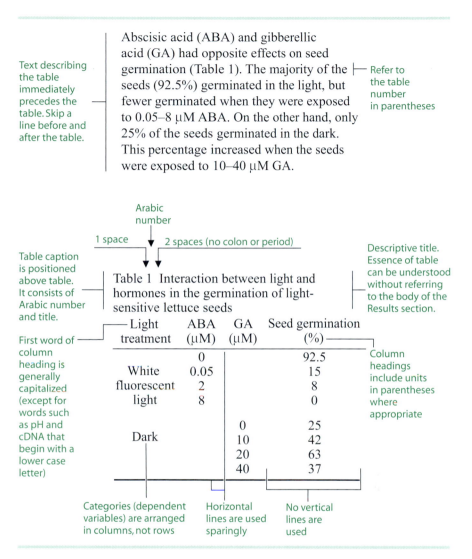

Figure 4.2 Excerpt from a Results section showing a properly formatted table preceded by the text that describes the data in the table.

From the table title alone, the reader should be able to understand the essence of the table without having to refer to the body (text) of the Results section. For simple tables, it may suffice to use a precise noun phrase rather than a full sentence for the title. For more complex tables, one or more full sentences may be required. Either way, English grammar rules apply:

- Do not capitalize common nouns (*general* classes of people, places, or things) unless they begin the phrase or sentence.
- Capitalize proper nouns (names of *specific* people, places, or things).
- Do not capitalize words that start with a lower case letter (for example, pH, mRNA, or cDNA), even if they begin a sentence.

Some examples of faulty and preferred titles are shown below.

FAULTY: Table 1 The Relationship Between Light and Hormones in the Germination of Light-Sensitive Lettuce Seeds

EXPLANATION: Do not capitalize common nouns.

FAULTY: Table 1 Table of interaction between light and hormones in the germination of light-sensitive lettuce seeds

EXPLANATION: Do not start a title with a description of the visual.

FAULTY: Table 1 Seed germination data

EXPLANATION: Do not write vague and undescriptive titles.

REVISION: Table 1 Interaction between light and hormones in the germination of light-sensitive lettuce seeds

A table is always positioned *after* the text in which you refer to it (see Figure 4.2). Refer to the table number in parentheses at the end of the first sentence in which you describe the table contents. That way the reader can refer to the table as you describe what you consider to be important.

In your laboratory report, it is not necessary to include a table when you already have a graph that shows the same data. Make *either* a table *or* a graph—not both—to present a given data set.

Tables can be constructed in either Microsoft Word (see "The Insert Tab" section in Appendix 1) or Microsoft Excel (see Figure A2.6 in Appendix 2).

Table preparation checklist
- ☐ Categories arranged in columns, not rows
- ☐ Column headings include units (where appropriate)
- ☐ Format correct (minimal lines)
- ☐ Table title descriptive
- ☐ Table title grammatically correct
- ☐ Table caption positioned above the table

Line graphs (XY graphs) and scatterplots

Line graphs display a relationship between **two or more quantitative variables**. What we hope to learn from the data in part determines the format of the XY graph (Figure 4.3). For example, in **observational studies**, scientists observe individuals and measure variables that they are interested in. Quite often, the purpose of an observational study is to look for a pattern in nature. Patterns may be easier to spot when the numerical data are plotted as a **scatterplot**, one kind of line graph in which the individual data points are not connected (Figure 4.3A). By convention, if one of the variables explains or influences the other, then this so-called explanatory or **independent variable** is plotted on the x-axis. The variable that shows the response, also called the **dependent variable**, is plotted on the y-axis. In some observational studies, there may not be a causative relationship between the two variables, in which case it doesn't matter which variable is plotted on which axis.

The first step in describing a pattern on a scatterplot is to look at the form and direction of the data. The **form** may be linear, curved, clustered, or random; because many relationships in nature are linear, exponential, or logarithmic, keep an eye out for these kinds of forms. The **direction** indicates whether the relationship between the variables is positive (large values for y correspond to large values for x and vice versa) or negative (large values for y correspond to small values for x and vice versa), or if there is no change in y with x or vice versa. Once you've described the form and direction of the scatterplot, try to assess the **strength** of the relationship. How closely do the points follow the form? A lot of scatter and the presence of outliers indicate a weak relationship.

Our eyes are pretty good at recognizing when the data fall on a straight line, but we need a more objective way to assess the strength of the relationship. One such indicator is called correlation (r), whose rather complex formula produces values between -1 and 1. Correlation values near 0 indicate a weak linear relationship, with the strength of the relationship increasing as r approaches -1 (when the direction is negative) or 1 (when the direction is positive). When the data in an observational study show a strong linear rela-

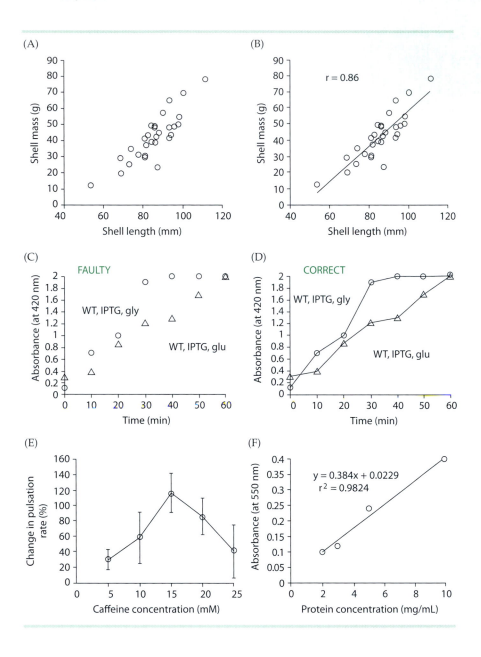

tionship, scientists may superimpose a straight line on the scatterplot and display *r* as a measure of the strength of the relationship (Figure 4.3B).

In observational studies, scientists measure a variable of interest without trying to influence the response. On the other hand, in **experiments**, scientists impose a treatment on individuals and then observe how the treatment

◀ **Figure 4.3** Different line graph formats showing relationships between variables. (A) A scatterplot displays numerical observations with the purpose of determining whether there is a relationship between shell mass and shell length. (B) A scatterplot with a straight line added shows that there is a strong linear relationship between the two variables. (C) The relationship between the independent and dependent variables in each treatment group is hard to see when the points are not connected. (D) The relationship is much easier to see when the points are connected with straight or smoothed lines. (E) Error bars show variability about the mean. (F) A least-squares regression line and its equation are used to predict one variable when the other is known. Regression lines are used only in specific situations when the linear relationship between the two variables is clearly established.

affects their responses. The purpose of an experiment, therefore, is to determine the effect of one variable (the explanatory or independent variable) on another (the response or dependent variable). By convention, the independent variable (the one the scientist manipulates) is plotted on the x-axis and the dependent variable (the one that changes in response to the independent variable) is plotted on the y-axis.

On a scatterplot, the individual data points are not connected, because the purpose of the graph is to determine the form, direction, and strength of the relationship between the variables. In contrast, in an experiment, scientists want to know **how the imposed treatments affect the response**. To make it easier to see this effect, **the data points are connected by straight or smoothed lines**. Lines avoid confusion particularly when there is more than one data set on a graph (compare Figures 4.3C and D). Never show the lines without the experimental data, however.

Data points displayed on graphs are typically a summary of the raw data, with each point representing the mean value calculated by averaging many replicates. **To show variability in the measured values** (especially when the data are distributed normally about the mean), authors may include **error bars** on their graphs (Figure 4.3E). An explanation of what the error bars represent—standard deviations or standard errors of the mean—should be given in the figure title along with the number of observations.

Finally, **standard curves** represent a special case of line graph, whose purpose is to predict one variable when the other is known. First the data points are plotted as a scatterplot, and then a least-squares regression line (best-fit line) is fitted to the points (Figure 4.3F). The square of the correlation, r^2, describes how well the regression line fits the data. The closer the r^2 value is to 1, the better the fit. The better the fit, the closer the predicted value will be to the true value of the unknown variable.

Figures are always numbered and titled *beneath* the visual (Figure 4.4; see also Figure 4.1). The captions may be centered or placed flush on the left margin of the report. Arabic numbers are used, and the figures are numbered consecutively in the order they are discussed in the text.

The percentage of male gametophytes increased with population density of the wild type strain, but there were no male gametophytes at any density in the Her 1 strain (Figure 2).

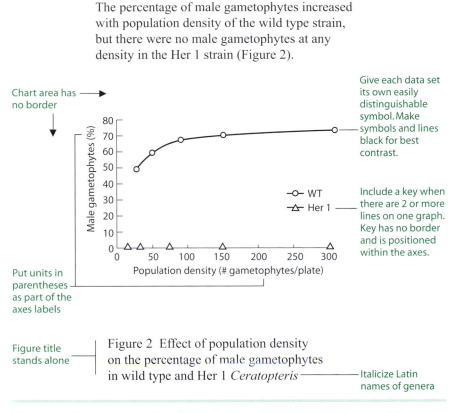

Chart area has no border

Give each data set its own easily distinguishable symbol. Make symbols and lines black for best contrast.

Include a key when there are 2 or more lines on one graph. Key has no border and is positioned within the axes.

Put units in parentheses as part of the axes labels

Figure title stands alone

Figure 2 Effect of population density on the percentage of male gametophytes in wild type and Her 1 *Ceratopteris*

Italicize Latin names of genera

Figure 4.4 Excerpt from a Results section showing a properly formatted figure with two sets of data. A legend (key) is needed to distinguish the two lines. The text that describes the figure precedes it.

From the figure title alone, the reader should be able to understand the essence of the figure without having to refer to the body (text) of the Results section. For simple figures, it may suffice to use a precise noun phrase rather than a full sentence for the title. For more complex figures, one or more full sentences may be required. Either way, English grammar rules apply: Do not capitalize common nouns (*general* classes of people, places, or things) unless they begin the phrase or sentence. Capitalize proper nouns (names of *specific* people, places, or things). Do not capitalize words that start with a lower case letter (for example, pH, mRNA, or cDNA), even if they begin a sentence. Some examples of faulty and preferred titles are shown here.

FAULTY: Figure 1 The Effect of Population Density on the Development of Male Gametophytes

EXPLANATION: Do not capitalize common nouns.

FAULTY: Figure 1 Percentage of male gametophytes vs. population density

EXPLANATION: Do not restate the y-axis label versus the x-axis label as the figure title.

FAULTY: Figure 1 shows the effect of population density on the percentage of male gametophytes in wild type *Ceratopteris*

EXPLANATION: Separate the figure number and the title.

FAULTY: Figure 1 Line graph of the effect of population density on the percentage of male gametophytes in wild type *Ceratopteris*

EXPLANATION: Do not start a title with a description of the visual.

FAULTY: Figure 1 Averaged class data for C-fern experiment

EXPLANATION: Do not write vague and undescriptive titles.

REVISION: Figure 1 Effect of population density on the percentage of male gametophytes in wild type *Ceratopteris*

If there is more than one data set (line) on the figure, you have three options:

- Add a brief label (no border, no arrows) next to each line.
- Use a different symbol for each line and label the symbols in a key (as in Figure 4.4). Place the key without a border within the axes of the graph. This is the easiest option if you are using Excel to plot your data.
- If the first two options make the figure look cluttered, identify the symbols in the figure caption.

All three formats are acceptable in scientific papers as long as you use them consistently.

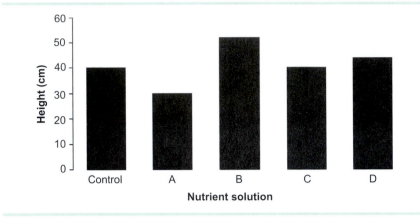

Figure 4.5 Final height of corn plants after 4-week treatment with different nutrient solutions. This figure is an example of a column graph.

Figures in your laboratory report should be prepared according to the guidelines specified by the Council of Science Editors (CSE Manual 2006). Although you may wish to plot a rough draft of your graphs by hand, you should learn how to use computer plotting software to make graphs. Microsoft Excel is a good plotting program for novices (see Appendix 2) because it is readily available and fairly easy to use. The time you invest now in learning to plot data on the computer will be invaluable in your upper-level courses and later in your career.

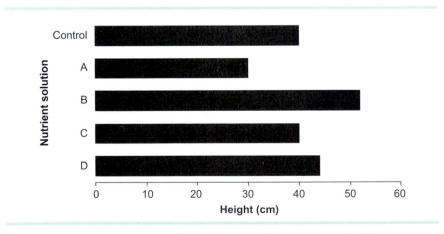

Figure 4.6 Final height of corn plants after 4-week treatment with different nutrient solutions. This figure is an example of a horizontal bar graph.

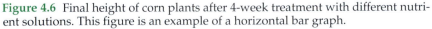

Bar graphs

A bar graph allows you to compare individual sets of data when **one of the variables is categorical** (not quantitative)—this is the main difference between line graphs and bar graphs. Bar graphs are more flexible than pie charts because any number of categories can be compared; the percentages do not have to total 100%. Error bars may be centered at the top of each data bar to show variability in the measured data. When the data bars are black, only half error bars are used.

Consider an experiment in which you want to compare the final height of the same species of plant treated with four different nutrient solutions. The nutrient solution is the non-numerical, categorical variable; the height is the response variable. The data bars can be arranged vertically (Figure 4.5) or horizontally (Figure 4.6). The arrangement you use often depends on your sub-discipline, but the horizontal arrangement is more practical when the category labels are long.

The bars should be placed sequentially, but if there is no particular order, then put the control treatment bar far left in column graphs or at the top in horizontal bar graphs. Order the experimental treatment bars from shortest to longest (or vice versa) to facilitate comparison among the different conditions. The baseline does not have to be visible, but all the bars must be aligned as if there were a baseline.

The bars should always be wider than the spaces between them. In a graph with clustered bars, make sure each bar has sufficient contrast so that it can be distinguished from its neighbor (Figure 4.7). Instructions for plotting bar graphs in Excel 2010 are given in Appendix 2.

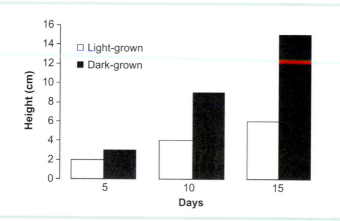

Figure 4.7 Difference in height of groups of light-grown and dark-grown bean seedlings at 5, 10, and 15 days after planting. This figure is an example of a clustered bar graph. Each bar in the cluster must be easily distinguishable from its neighbor.

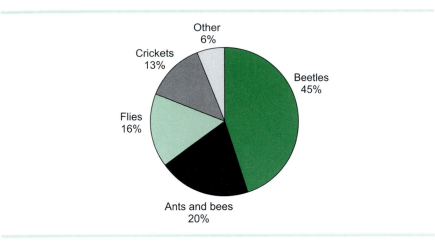

Figure 4.8 Composition of insects in backyard survey. Pie graphs are used to show data as a percentage of the total data.

Pie graphs

A pie graph is used to show data as a percentage of the total data. For example, if you were doing a survey of insects found in your backyard, a pie graph would be effective in showing the percentage of each kind of insect out of all the insects sampled (Figure 4.8). There should be between 2 and 8 segments in the pie. Place the largest segment in the right-hand quadrant with the segments decreasing in size clockwise. Combine small segments under the heading "Other." Position labels and percentages horizontally outside of the segments for easy reference. Instructions for plotting pie charts in Excel 2010 are given in Appendix 2.

> **Figure preparation checklist**
> ☐ Right type of graph
> ☐ Format correct (symbols, lines, legend, axis scale, no gridlines, no border)
> ☐ Figure title descriptive
> ☐ Figure title grammatically correct
> ☐ Figure caption positioned below the figure

Organizing your data

Reread the laboratory exercise to see if your instructor has provided specific guidelines on the kinds of visuals to include in the Results section. If you have to make the decision on your own, ask yourself the following questions:

- Can I state the results in one sentence? If so, then **no visual** is needed.
- Are the numbers themselves more important than the trend shown by the numbers? If so, then use a **table**.
- Is the trend more important than the numbers themselves? If so, use a **graph**.
  - Are both variables quantitative? If so, then use a **line graph**.
  - Is one of the variables categorical (not quantitative)? If so, then use a **bar graph**.
- Are the results descriptive rather than quantitative? If so, use **photos and images**.

The following examples demonstrate that there may be more than one good way to organize the data. Some visuals may be more appropriate than others, and in some cases, no visual may be the best alternative.

EXAMPLE **1:**   *Brassica* seeds were placed on filter paper saturated with pH 1, 2, 3, or 4 buffered solutions. The positive control was filter paper saturated with water. After 2 days, 100% of the seeds in the positive control germinated. No seeds germinated in any of the buffered solutions.

POSSIBLE SOLUTION: **No visual** is needed because the results can be summarized in one sentence: "After 2 days, 100% of the seeds that imbibed water germinated, but none of the seeds that were treated with buffered solutions pH 1, 2, 3, or 4 germinated."

EXAMPLE **2:**   Light-sensitive lettuce seeds placed on filter paper saturated with water were exposed to the same fluence of white fluorescent, red, far-red, green, and blue light treatments as well as darkness, and the percentage that germinated was determined 30 hours later.

POSSIBLE SOLUTION: Since some of the data are categorical rather than quantitative (colors rather than wavelengths of light), either a *table* or a *bar graph* (Figure 4.9) works well to display the results.

(A)

Table 1 Effect of light treatment on percentage of light-sensitive lettuce seeds germinated after 30 hr

Light treatment	Seed germination (%)
Red	76
White	65
Blue	44
Far-red	38
Green	37
Dark	30

(B)

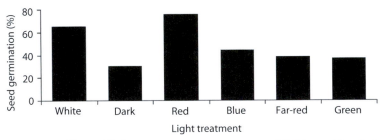

Figure 1 Effect of light treatment on percentage of light-sensitive lettuce seeds germinated after 30 hr

Figure 4.9 Example 2 data summarized in (A) a table or (B) a bar graph. The positive and negative controls are placed to the left, and, if there is no particular order to the categories (colors in this example), arrange the bars in order of the longest to the shortest (or vice versa).

INAPPROPRIATE SOLUTION: A *line graph* is not appropriate because both variables are not quantitative. We only know the colors of light, not the exact wavelengths. *Text only* is **not** appropriate because listing the seed germination percentages in a sentence is tedious to read and hard to comprehend.

EXAMPLE 3: The activity of an enzyme (catalase) was monitored at nine different temperatures in order to determine the optimal temperature for maximum activity.

POSSIBLE SOLUTION: If your emphasis is on the actual numbers rather than the trend, then display the results in a *table*. If the trend is more important than the numbers, use a *line graph and connect the points* with smoothed or straight lines (Figure 4.10).

(A)

Table 1 Effect of temperature on catalase activity

Temperature (°C)	Catalase activity (units of product formed · sec⁻¹)
4	0.039
15	0.073
23	0.077
30	0.096
37	0.082
50	0.040
60	0.007
70	0
100	0

(B)

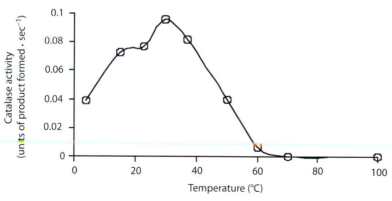

Figure 1 Effect of temperature on catalase activity

Figure 4.10 Example 3 data summarized in (A) a table or (B) a line graph. When data are summarized in a table, the emphasis is placed on the numbers rather than on the trend. For the same data, a line graph is more effective than a table in showing the trend.

Make Connections

Now that the "meat" of your report is done, it's time to describe how your work fits into the existing body of knowledge. These connections are made in the Discussion and Introduction sections.

Write the discussion

The Discussion section gives you the opportunity to **interpret your results, relate them to published findings, and explain why they are important**. The structure of the discussion is specific to broad, as illustrated by the following triangle.

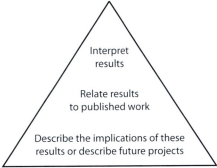

Interpret
results

Relate results
to published work

Describe the implications of these
results or describe future projects

Tense. When *describing* your own results, use *past* tense. However, when you use scientific fact to *explain* your results, use *present* tense.

PAST TENSE: The optimal temperature for catalase activity *was* 30°C (Figure 1).

EXPLANATION: Past tense signifies that this statement is limited to your own results.

PRESENT TENSE: Enzyme activity *is* low at cold temperatures because of decreased kinetic energy. At high temperatures, however, enzyme activity *is* low because the enzymes *are* denatured.

EXPLANATION: Present tense signifies that these statements are generally valid and considered to be scientific fact.

Voice. Use primarily active voice to make your writing clear and dynamic.

Style. When explaining your results, never use the word *prove*. Instead, use words and phrases like *provide evidence for*, *support*, *indicate*, *demonstrate*, or *strongly suggest*. The reason for this choice of words lies in the logic behind the scientific method. If our results match our predictions, then there is evidence that our hypotheses are correct. When many scientists get the same results independently, then the support for a given hypothesis grows. Scientists are reluctant to use the word "prove," because there is always a chance that a future study may provide conflicting evidence.

> **FAULTY:**   These results prove that catalase was denatured at temperatures above 60°C.

> **REVISION:**   These results strongly suggest that catalase was denatured at temperatures above 60°C.

Organization. Start the discussion by **restating the objectives** of the current work. Then **recap each result** and **try to explain it**. Interpret the results so that the reader understands how you arrived at your conclusions. Ideally, there will be a one-to-one correspondence between the important points you address in the Discussion section with the problems or questions you stated in the Introduction section.

Especially in introductory biology labs, the results may not always work out the way we expect. **If your results defy explanation**, consider these possible reasons:

- Human error, including failure to follow the procedure, failure to use the equipment properly, failure to prepare solutions correctly, variability when multiple lab partners measure the same thing, and simple arithmetic errors. If you suspect that human error may have affected your results, then acknowledge its contribution.
- Numerical values were entered incorrectly in the computer plotting program.
- Sample size was too small.
- Variability was too great to draw any conclusions.

If you can rule out these possibilities, discuss your results with your lab partner, teaching assistant, or instructor. If there is an obvious error, an "outsider" may be able to spot it immediately. In any event, having a discussion with a knowledgeable individual may help you understand the concept, even if you obtained unexpected results.

Next, **compare your results with those in the literature**. Do your findings support or contradict results published previously? Did you use a different method, but still obtain the same result? Or could a different method account for a conflicting result? If there are inconsistencies in your data, point them out. Discuss possible sources of error.

Read the Discussion section of published papers to learn how scientists present convincing arguments for their conclusions. Always paraphrase information you obtained from others and cite the source (see "Documenting Sources" on pp. 81–94).

Finally, if warranted, **add a conclusion about the significance of your work**. You may describe how your results apply to a related field or possible future work that focuses on an interesting observation.

Write the introduction

After having written drafts of the Materials and Methods, Results, and Discussion sections, you should be intimately familiar with the procedure, the results, and what the results mean. Now you are in a position to put your investigation into perspective. What was already known about the topic? Were there any unanswered questions? Why did you carry out this investigation?

The structure of the introduction is broad to specific, just the opposite of that of the discussion.

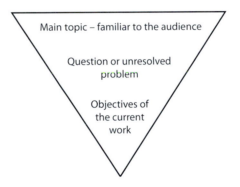

Main topic – familiar to the audience

Question or unresolved problem

Objectives of the current work

Organization. The introduction consists of two main parts:

- Background information from the literature, and
- Objectives of the current work.

The opening sentence of the Introduction section is usually a **general observation or result** familiar to readers in that discipline. Subsequent sentences **narrow down the topic** to the specific focus of the current study. Subsequent paragraphs then provide background information from the literature and **describe unanswered questions** or inconsistencies. The **objectives of the current work** are usually stated in the last paragraph of the Introduction section.

Tense. In the course of providing background information on your topic, you will discuss scientific fact that is based on findings published in research papers. When describing scientific fact, use *present* tense. When stating the objectives of your study, use *past* tense. Past tense is preferred because proposing objectives is a completed action that you carried out before starting your actual study.

Voice. Active voice is preferred because it makes sentences shorter and more direct.

Effective Advertising

The whole point of writing your paper is to communicate your work to your fellow scientists. The abstract and the title are the primary tools potential readers will use to decide whether or not they are interested in your work.

Write the abstract

The abstract is a **summary of the entire paper** in 250 words or less. It contains:

- An introduction (scope and purpose)
- A short description of the methods
- The results
- Your conclusions

There are no literature citations or references to figures in the abstract.

After the title, the abstract is the most important part of the scientific paper used by readers to determine initial interest in the author's work. Abstracts are indexed in databases that catalogue the literature in the biological sciences. If an abstract suggests that the author's work may be relevant to your own work, you will probably want to read the whole article. On the other hand, if an abstract is vague or essential information is missing, you will probably decide that the paper is not worth reading. When you write the abstract for your own laboratory report, put yourself in the position of the reader. If you want the reader to be interested in your work, write an effective abstract.

Writing the abstract is difficult because you have to condense your entire paper into 250 words or less. One strategy for doing this is to list the key points of each section, as though you were taking notes on your own paper. Then write the key points in full sentences. Revise the draft for clarity and conciseness using strategies such as using active voice, combining choppy sentences with connecting words, rewording run-on sentences, and eliminating redundancy. With each revision, look for ways to shorten the text so that the resulting abstract is a concise and accurate summary of your work.

The ability to write abstracts is important to a scientist's career. Should you someday wish to present your research at an academic society meeting, such as the Society for Neuroscience, the American Association for the Advancement of Science, or the National Association of Biology Teachers (to name just a few), you will be asked to submit an abstract of your presentation to the committee in charge of the meeting program. Your chances of being among the select field of presenters at these meetings are much better if you have learned to write a clear and intelligent abstract.

Write the title

The title is a **short, informative description** of the essence of the paper. You may choose a working title when you begin to write your paper, but revise the title after subsequent drafts. Remember that readers use the title to determine initial interest in the paper, so descriptive accuracy is the most essential element of your title. Brevity is nice if it can be achieved. Some journals (especially the British ones) are fond of puns and humor in their titles, but this kind of thing may be better left for later in your career.

Here are some examples of vague and undescriptive titles:

FAULTY: Quantitative Protein Analysis

FAULTY: The Assessment of Protein Content in an Unknown Sample

FAULTY: Egg White Protein Analysis

EXPLANATION: These titles leave the reader wondering what method of protein analysis was used and what sample was analyzed.

REVISION: Assessment of protein concentration in egg white using the biuret method

Here is another series of examples in which adding specific details improves the title from the first to the last example:

FAULTY: Study of an Enzymatic Reaction

EXPLANATION: Specify the variables you studied. Specify the enzyme and the substrate in the reaction.

FAULTY: Initial velocity of enzymatic reactions under varying conditions

EXPLANATION:	Was *more than one* enzymatic reaction studied? What were the *specific conditions*? If you only studied one reaction, use the singular.
FAULTY:	Study of 3 factors on an enzymatic reaction
EXPLANATION:	What were the *specific factors*? Specify the enzyme and the substrate in the reaction.
FAULTY:	Enzyme and substrate concentration and its effects on initial velocity as seen with peroxidase and hydrogen peroxide
EXPLANATION:	The variables, the enzyme, and the substrate are present, but the sentence structure is a little awkward.
REVISION:	Effect of substrate and enzyme concentration and hydroxylamine (an inhibitor) on the initial velocity of the peroxidase–hydrogen peroxide reaction

Documenting Sources

Whenever you use another person's ideas, whether they are published or not, you must document the source. This is done by citing the source in an abbreviated form in the text (**in-text citation**) and then giving the full reference in the References section at the end of the paper (**end reference**). An exception to this practice is personal communications, which are cited in the text, but are not listed among the end references. Only sources that have been cited in the text may be included in the References section.

The CSE Manual (2006) recommends using the Citation-Sequence System (C-S), the Citation-Name System (C-N), or the Name-Year System (N-Y) for documenting your sources. The system you actually use depends on your instructor's preference or on the format specified by the particular scientific journal in which you aspire to publish.

In all three systems, the *in-text citation* is intended to be inconspicuous. A superscripted number or a number in parentheses (C-S and C-N systems) or authors' names and year (N-Y system) are minimally disruptive to the flow of the sentence. Contrast this style with the lengthy introduction practiced in some disciplines in the humanities: "According to Warne and Hickock in their 1989 paper published in *Plant Physiology*, antheridiogen may be related to the gibberellins." **Do not use this style in your lab reports!**

With regard to the *end reference*, the systems differ in the sequence of information and the listing of the month of publication. In the N-Y system, the year of publication follows the author's name; in the C-S and C-N systems, the year follows the journal name. The month of publication is only used in the C-S and C-N systems.

The Name-Year system has the advantage that people working in the field will know the literature and, on seeing the authors' names, will understand the in-text citation without having to check the end reference. With the Citation-Sequence and Citation-Name systems, the reader must turn to the reference list at the end of the paper to gain the same information. Regardless of which system you use, learn the proper way to format both the in-text citation and the end reference and use one system consistently throughout any given paper.

Finally, do not list sources in the end reference list, which you personally have not seen. If you feel that the original source is important enough to be cited, use the following approach:

Author (year) as cited by Author (year)

The Name-Year system

The *in-text citation* consists of author(s) and year. The author(s) may be cited in parentheses at the end of the sentence or they may be the subject of the sentence, as shown in the following examples:

TABLE 4.4   Number of authors determines how the authors are cited in N-Y system

Number of Authors	Author as Subject	Parenthetical Reference (The comma between author[s] and year is optional.)
1	Author's last name (year) found that…	(Author's last name, year)
2	First author's last name and second author's last name (year) found that …	(First author's last name and second author's last name, year)
3 or more	First author's last name followed by and others or *et al.* (year) found that …	(First author's last name and others, year) or *et al.* instead of *and others*

Note: If you cite more than one paper published by the same author in *different* years, list them in chronological order: (Dawson 2001, 2003). If you cite more than one paper published by the same author in the *same* year, add a letter after the year: "…was described in recent work by Dawson (1999a, 1999b)."

PARENTHESES:	C-fern gametophytes respond to antheridiogen only for a short time after inoculation (Banks and others 1993).
AS THE SUBJECT:	Banks and others (1993) found that C-fern gametophytes respond to antheridiogen only for a short time after inoculation.

The **number of authors** determines how the *in-text citation* is written in the N-Y system (Table 4.4). For *one* author, write the author's last name and year. For *two* authors, write both authors' last names separated by the word *and* followed by the year. For *three or more* authors, write the first author's last name, the words *and others* (or *et al.*), and then the year.

In the *end references*, the sources are listed in **alphabetical order** according to the first author's last name. The format of the source determines which elements are included (Table 4.5). When there are 10 or fewer authors, list all authors' names. When there are more than 10 authors, list the first 10 and then write *et al.* or *and others* after the tenth name. Write each author's name in the form Last name First initials. Use a comma to separate one author's name from the next. Use a period only after the last author's name.

Examples of in-text citations and their corresponding end references are given in Table 4.6.

The Citation-Sequence system

The *in-text citation* consists of a superscripted endnote (never a footnote) or a number in parentheses or square brackets within or at the end of the paraphrased sentence. The first reference cited is number 1, the second reference cited is number 2, and so on.

SUPERSCRIPTED ENDNOTE:	There are four commonly used methods for determining protein concentration: the biuret method[1], the Lowry method[2], the Coomassie Blue (CB) dye-binding method[3], and the bicinchoninic acid (BCA) assay[4].
PARENTHESES:	The Kjeldahl procedure is time-consuming and requires a large amount of sample (1, 2).
BRACKETS:	Several review articles compare the advantages and disadvantages of these protein assays [5–10].

In the *end references*, the sources are listed in **numerical order** (in the order of citation). The format of the source determines which elements are

TABLE 4.5	General format of two systems of source documentation used in scientific papers

N-Y End Reference System

The references are listed in **alphabetical order**. The last name is written first, followed by the initials. When there are 10 or fewer authors, list all authors' names. When there are more than 10 authors, list the first 10 and then write *et al.* or *and others* after the tenth name. Type references with hanging indent format.

Journal article	First author's last name First initials, Subsequent authors' names separated by commas. Year of publication. Article title. Journal title Volume number(issue number): inclusive pages.
Article in book	First author's last name First initials, Subsequent authors' names separated by commas. Year of publication. Article title. In: Editors' names followed by a comma and the word *editors*. Book title, edition. Place of Publication: Publisher. pp inclusive pages.
Book	First author's or editor's last name First initials, Subsequent authors' or editors' names separated by commas. Year of publication. Title of book. Place of Publication: Publisher. Total number of pages in book followed by *p*.

C-S End Reference System

The references are listed **in the order they are cited**. The author's last name is written first, followed by the initials. When there are 10 or fewer authors, list all authors' names. When there are more than 10 authors, list the first 10 and then write *et al.* or *and others* after the tenth name.

Journal article	Number of the citation. First author's last name First initials, Subsequent authors' names separated by commas. Article title. Journal title Year Month; Volume number (issue number): inclusive pages.
Article in book	Number of the citation. First author's last name First initials, Subsequent authors' names separated by commas. Article title. In: Editors' names followed by a comma and the word *editors*. Book title, edition. Place of Publication: Publisher; Year of publication. pp inclusive pages.
Book	Number of the citation. First author's or editor's last name First initials, Subsequent authors' or editors' names separated by commas. Title of book. Place of Publication: Publisher; Year of publication. Total number of pages in book followed by *p*.

included (Table 4.5). When there are 10 or fewer authors, list all authors' names. When there are more than 10 authors, list the first 10 and then write *et al.* or *and others* after the tenth name. Write each author's name in the form Last name First initials. Use a comma to separate one author's name from the next. Use a period only after the last author's name.

Examples of in-text citations and their corresponding end references are given in Table 4.6.

The Citation-Name system

In the *end references*, the sources are listed in **alphabetical order** according to the first author's last name. The year and month of publication follow the journal name, as in C-S end reference format. The references are then numbered sequentially so that the first reference is number 1, the second reference is number 2, and so on. The *in-text citations* consist of superscripted endnotes (never footnotes) or a number in parentheses or square brackets within or at the end of the paraphrased sentence.

Unpublished laboratory exercise

Unpublished material is usually not included in the References section. However, if your instructor asks that you cite laboratory exercises in your laboratory report, the *end reference* could look like this:

C-S: #. Author (omit if unknown). Title of lab exercise. Course number, Department, University. Year.

N-Y: Author (if unknown, replace with title of lab exercise). Year. Title of lab exercise. Course number, Department, University.

In N-Y format, the *in-text citation* for an unpublished lab exercise would include the author(s) and year, or, if the author is unknown, the title and year. The use of anonymous is not recommended (CSE Manual 2006).

Personal communication

Unpublished information obtained during a discussion or by attending a lecture should be acknowledged when you use it in your lab report or scientific paper. The in-text citation includes the authority, the date, and the words "personal communication" or "unreferenced." For example:

…may be explained by possible contamination from viruses or bacteria (M. Pizzorno, personal communication, 2012 Oct 30).

It is **not necessary** to include personal communications in the references.

Reference Formats

TABLE 4.6	Examples of in-text citation and end reference format of two systems of source documentation used in scientific papers

Name-Year System

IN-TEXT CITATIONS

1 author 3 or more authors 2 authors 1 author	Gametophytes of the tropical fern *Ceratopteris richardii* (C-fern) develop either as males or hermaphrodites. Their fate is determined by the pheromone antheridiogen (Näf 1979; Näf and others 1975). Banks and others (1993) found that gametophytes respond to antheridiogen only for a short time between 3 and 4 days after inoculation. Although the structure of antheridiogen is unknown, it is thought to be related to the gibberellins (Warne and Hickok 1989). Gibberellins are a group of plant hormones involved in stem elongation, seed germination, flowering, and fruit development (Treshow 1970).

CORRESPONDING END REFERENCES

Journal article	Banks J, Webb M, Hickok L. 1993. Programming of sexual phenotype in the homosporous fern *Ceratopteris richardii*. Inter. J. Plant Sci. 154(4): 522-534.
Article in book	Näf U. 1979. Antheridiogens and antheridial development. In: Dyer AF, editor. The Experimental Biology of Ferns. London: Academic Press. pp. 436-470.
Journal article	Näf U, Nakanishi K, Endo M. 1975. On the physiology and chemistry of fern antheridiogens. Bot. Rev. 41(3): 315-359.
Book	Treshow M. 1970. Environment and Plant Response. New York: McGraw-Hill. 250 p.
Journal article	Warne T, Hickok L. 1989. Evidence for a gibberellin biosynthetic origin of *Ceratopteris* antheridiogen. Plant Physiol. 89(2): 535-538.

TABLE 4.6 *Continued*	
	Citation-Sequence System

	IN-TEXT CITATIONS
Sources are listed in the order they are cited	Gametophytes of the tropical fern *Ceratopteris richardii* (C-fern) develop either as males or hermaphrodites. Their fate is determined by the pheromone antheridiogen (1, 2). Gametophytes respond to antheridiogen only for a short time between 3 and 4 days after inoculation (3). Although the structure of antheridiogen is unknown, it is thought to be related to the gibberellins (4). Gibberellins are a group of plant hormones involved in stem elongation, seed germination, flowering, and fruit development (5).

	CORRESPONDING END REFERENCES	
Article in book	1.	Näf U. Antheridiogens and antheridial development. In: Dyer AF, editor. The Experimental Biology of Ferns. London: Academic Press; 1979. pp. 436-470.
Journal article	2.	Näf U, Nakanishi K, Endo M. On the physiology and chemistry of fern antheridiogens. Bot. Rev. 1975 Jul-Sep; 41(3): 315-359.
Journal article	3.	Banks J, Webb M, Hickok L. Programming of sexual phenotype in the homosporous fern *Ceratopteris richardii*. Inter. J. Plant Sci. 1993 Dec; 154(4): 522-534.
Journal article	4.	Warne T, Hickok L. Evidence for a gibberellin biosynthetic origin of *Ceratopteris* antheridiogen. Plant Physiol. 1989 Feb; 89(2): 535-538
Book	5.	Treshow M. Environment and Plant Response. New York: McGraw-Hill; 1970. 250 p.

Book references give the total number of pages in the book, not the pages from which you extracted the information

Internet sources

In the previous section you learned that the in-text citation and end reference format differs for journal articles, articles in a book, and books. These differences apply to both print and online publications. For a journal article, therefore, you should be able to locate on the website the names of the authors, a title, the journal name, a date of publication, the volume and issue number, and the extent (number of pages or similar). Besides this basic information, the CSE Manual (2006) recommends that you provide two additional items when your reference comes from the Internet: the URL (uniform resource locator) and the date accessed. For your lab reports, it is sufficient to treat references obtained online as print sources (unless your instructor tells you otherwise). If you would like to publish your work in a journal that adheres strictly to CSE guidelines, however, the following section shows the in-text citation and end reference format for an online journal article. For a comprehensive discussion of Internet citation formats along with many examples, see Patrias (2007).

> When URLs are used in text, they are enclosed in angle brackets (< >) to distinguish them from the rest of the text. (URLs can also be printed in color, in which case the angle brackets are not required.) Every character in a URL is significant, as are spaces and capitalization. Very long URLs can be broken before a punctuation mark (tilde ~, hyphen -, underscore _ , period ., forward slash /, backslash \, or pipe |). The punctuation mark is then moved to the next line.

Journal articles

The *in-text citation* for an online journal article is exactly the same as that for a printed journal article (see Table 4.6). A good approach for writing the *end reference* of an online journal article is to first locate the information you would need for a printed journal article, and then add the Internet-specific items (CSE Manual 2006). Choose one of the three systems—Name-Year, Citation-Sequence, or Citation-Name—and position the elements accordingly.

The general format for a *printed* end reference in the **Name-Year system**, including punctuation, is:

> Author(s). Date of publication. Title of article. Title of journal plus
>   volume(issue): Inclusive page numbers

The corresponding format for an *online* reference, with the Internet information shown here in bold, is:

Author(s). Date of publication. Title of article. Title of journal **[Internet]. [date updated; date cited]**; Volume(issue): Inclusive page numbers. **Available from: URL**

A screen shot of an online journal article web page is shown in Figure 4.11 and the elements required for citation are labeled. The corresponding end reference in **Name-Year format** with Internet-specific information is as follows:

Wariishi H, Nonaka D, Johjima T, Nakamura N, Naruta Y, Kubo S, Fukuyama K. 2000. Direct binding of hydroxylamine to the heme iron of *Arthromyces ramosus* peroxidase. Substrate analogue that inhibits compound I formation in a competitive manner. J Biol Chem [Internet]. [cited 2012 Oct 29]; 275(42): 32919–32924. Available from: http://www.jbc.org/content/275/42/32919.long

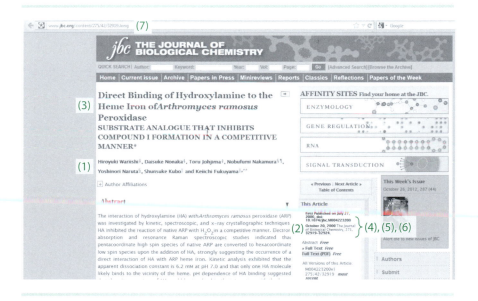

Figure 4.11 Web page for an online journal article. The basic information needed to cite a journal article includes (1) authors, (2) date of publication, (3) article title, (4) journal title, (5) volume and issue number (if given), and (6) inclusive pages. In addition, for an online journal article, include (7) the URL and the date accessed in the end reference.

The general format for a *printed* reference in the **Citation-Sequence** system, including punctuation, is:

> Number of the citation. Author(s). Title of article. Title of journal plus year and month; Volume(issue): Inclusive page numbers

The corresponding format for an *online* reference, with the Internet information shown here in bold, is:

> Number of the citation. Author(s). Title of article. Title of journal **[Internet]**. Year and month **[date updated; date cited]**; Volume(issue): Inclusive page numbers. **Available from: URL**

The end reference for the same online journal article shown in Figure 4.11 in **Citation-Sequence format** would be:

> 1. Wariishi H, Nonaka D, Johjima T, Nakamura N, Naruta Y, Kubo S, Fukuyama K. Direct binding of hydroxylamine to the heme iron of *Arthromyces ramosus* peroxidase. Substrate analogue that inhibits compound I formation in a competitive manner. J Biol Chem [Internet]. 2000 Oct [cited 2012 Oct 29]; 275(42): 32919–32924. Available from: http://www.jbc.org/content/275/42/32919.long

Databases

A database is a collection of records with a standard format. Databases may be text-oriented or numerical and their content is usually accessed by means of a search box. You may cite an entire database if your goal is to make the reader aware of its existence, or you may only cite a part of the database to document an individual record. Some databases are available on paper and CD-ROM as well as on the Internet. Specify the medium, as Internet databases may contain more recent information than the corresponding paper or CD-ROM versions.

The general format for citing a database in the **Name-Year system** is:

> Title of Database [medium designator]. Beginning date – ending date (if given). Edition. Place of Publication: Publisher. [date updated; date cited]. Available from: URL

To cite a database in the **Citation-Sequence system**, move the date after the publisher:

> Number of the citation. Title of Database [Medium Designator]. Edition. Place of Publication: Publisher. Beginning date – ending date (if given). [date updated; date cited]. Available from: URL

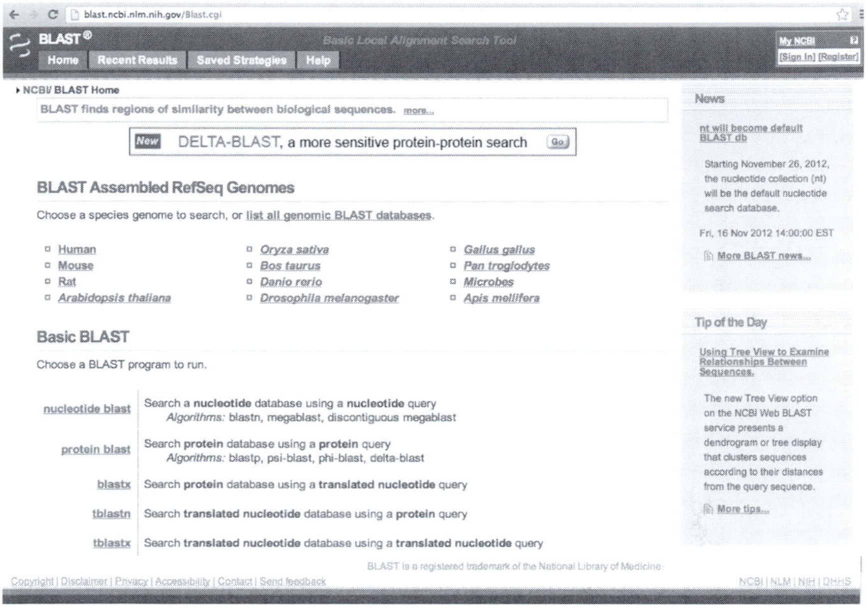

Figure 4.12 Homepage for the National Center for Biotechnology Information's BLAST database. To search for a specific nucleotide sequence, use **nucleotide blast**, one of the databases within the BLAST database.

As an example, the homepage of the BLAST database is shown in Figure 4.12. The nucleotide blast, protein blast, blastx, tblastn, and tblastx databases are individual websites within the larger BLAST website. When citing websites within websites, the following rule applies: Always cite the most specific organizational entity that you can identify (Patrias 2007). Database titles do not always follow the rules of English grammar and punctuation. Because they are proper nouns, however, reproduce the title as closely as possible to the format on the screen (maintain upper or lower case letters, run-together words, etc.). Sometimes the information needed for the reference may be absent or hard to find. In this example, the beginning to ending dates and the edition of the database are not specified. The location of the place of publication and the publisher are not given on this page, but can be found by clicking the **Contact** link at the bottom of the page. Do your best to reference the source with the information provided.

A good faith attempt at citing the nucleotide blast database in **Name-Year format** would be as follows:

> nucleotide blast [database on the Internet]. Bethesda (MD): National Library of Medicine (US), National Center for Biotechnology Information. [cited 2012 Oct 29]. Available from: http://blast.ncbi.nlm.nih.gov/Blast.cgi

In Citation-Sequence format:

> 1. nucleotide blast [database on the Internet]. Bethesda (MD): National Library of Medicine (US), National Center for Biotechnology Information. [cited 2012 Oct 29]. Available from: http://blast.ncbi.nlm.nih.gov/Blast.cgi

The *in-text citation* for a database in **Name-Year format** follows the same principles used for print publications (see Table 4.6) with a minor modification. The author is replaced with the title of the database and, when the date of publication is not known (as in the current example), the order of preference is the copyright date; the date of modification, update, or revision; and the date cited (CSE Manual 2006). An example of an *in-text citation* in **Name-Year format** for this database would be:

> There was a 100% match between the DNA sequence of Sample 1 and the SV40 sequence in the NCBI databank (nucleotide blast database, 2012).

Homepages

A homepage is the main page of a website, which provides links to different content areas of the site. Most of the information required to cite a website is found on the homepage. Make sure the organization or individual responsible for the website is reputable and, if possible, confirm information on the site using another source. The general format for citing a homepage in **Name-Year format** is:

> Title of Homepage [Internet]. Date of publication. Edition. Place of publication: publisher; [date updated; date cited]. Available from: URL

To cite a homepage in **Citation-Sequence format**, move the date after the publisher:

> Number of the citation. Title of Homepage [Internet]. Edition. Place of publication: publisher; date of publication [date updated; date cited]. Available from: URL

An example of a homepage is shown in Figure 4.13. All of the information required to cite this source is readily located. When the date of publication is not specified, the order of preference is the copyright date; the date of modification, update, or revision; and the date cited (CSE Manual 2006). In this example, the copyright date, preceded by a lower case *c* is used in the end reference. The end reference in **Name-Year format**:

> Stem Cells at the National Academies [Internet]. c2009. Washington DC: National Academy of Sciences; [cited 2012 Oct 29]. Available from: http://dels-old.nas.edu/bls/stemcells/what-is-a-stem-cell.shtml

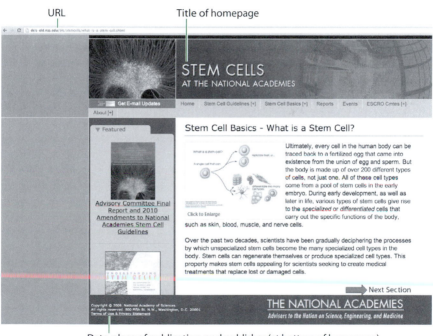

Figure 4.13 The National Academies' website for information on stem cells. Well-constructed homepages make it easy to find the title, date, responsible organization, and place of publication.

The end reference in Citation-Sequence format:

1. Stem Cells at the National Academies [Internet]. Washington DC: National Academy of Sciences; c2009 [cited 2012 Oct 29]. Available from: http://dels-old.nas.edu/bls/stemcells/what-is-a-stem-cell.shtml

The *in-text citation* for a homepage in **Name-Year format** follows the same principles used for print publications (see Table 4.6) with a minor modification. The author is replaced with the title of the homepage. For the year, the order of preference is the date of publication; the copyright date; the date of modification, update, or revision; and the date cited (CSE Manual 2006).

Emails and discussion lists

Electronic mail (email) and discussion lists (LISTSERVs, news groups, bulletin boards, etc.) are usually considered to be a form of personal communication (see p. 85). Information obtained through personal communication is cited in the in-text citation, but not in the end reference list. The in-text citation should include the name of the authority, the date, and the words "personal communication" or "unreferenced" to indicate that the citation is not listed in the References section.

Chapter 5

REVISION

Revision—reading your paper and making corrections and improvements—is an important task that usually does not get nearly the attention it deserves. Too many students write the first draft of their laboratory report the night before it is due and hand in the hard copy, still warm from the printer, without even having proofread it.

The truth is, most writers cannot produce a clear, concise, and error-free product on the first try. It may take several revisions before the writer is satisfied that he or she has conveyed, with clarity and logic, the motivation for writing the paper, the important findings, and the conclusions. Do not try to write and revise your entire paper in one "marathon" session. Instead, **break up the writing process** into multiple, shorter segments. The breaks give your mind time to process what you've written, get help if necessary, get feedback from your peer reviewer, and make final revisions.

Most excellent writers were not born that way. They achieved excellence through "deliberate practice" (Martin 2011). The old adage "practice makes perfect" applies not just to musicians and athletes, but to *you* as an aspiring author in the biological sciences. So if writing doesn't come naturally to you, take heart. Writing laboratory reports becomes easier with practice, especially if you learn from your mistakes.

Getting Ready to Revise

Take a break

The first step in revision is *not* to do it immediately after you have completed the first draft. You need to distance yourself from the paper to gain the objectivity needed to read the paper critically. So take a break, and go for a run or get a good night's sleep.

Slow down and concentrate

Find a quiet room where you won't be disturbed. Don't read your paper the same way you wrote it. Instead, **change its appearance** by increasing the zoom level or converting the Word document to a PDF (LR Communication Systems 1999; Corbett 2011). Even better, read a printed copy. Next, use some of your other senses to force yourself to **slow down**. For example, reading aloud involves the sense of hearing. Pointing to each word with your finger adds the sense of touch. If something doesn't sound right, trust your instincts. Figure out what is bothering you and fix the troublesome passage. Finally, **don't try to fix every kind of error in one pass**. If you do, you're sure to miss some.

Think of your audience

The rest of this chapter describes a systematic approach to revising your writing, whether for a lab report, a poster, or an oral presentation. Remember for whom you are writing and keep in mind the needs and motivations of your audience. Revise your writing with the goal of meeting or exceeding their expectations in a style that eases comprehension.

Editing

Revision can be divided into two stages: editing and proofreading. **Editing** is done first and involves reading for content and organization. The editing process proceeds from the broad to the specific. First, evaluate the overall structure of your paper. Then, read each individual section, paragraph, sentence, and word critically. And don't forget to check that the data were plotted correctly and that the description of each visual in the text is accurate. After editing is completed, **proofread** the paper. This involves correcting errors in spelling, punctuation, grammar, and overall format. The following sections provide specific guidance on what to look for at each step in the revision process. The "Laboratory Report Mistakes" section on pp. 138–141 and on <http://sites.sinauer.com/Knisely4e> lists common mistakes to look out for when you revise your lab report.

Evaluate the overall structure

If your instructor provided a rubric for your lab report, use it as a checklist for content and organization. You may also wish to print out the "Biology Lab Report Checklist" on pp. 127–128 from <http://sites.sinauer.com/Knisely4e>. If you are preparing a paper for submission to a journal or conference, follow the relevant *Instructions to Authors*.

TABLE 5.1 Checklist for section content	
Section	**Content**
Title	Contains keywords that describe the essence of your study
	Avoid "filler" phrases like "Analysis of the…," or "Study of the…"
Abstract	Contains an introduction, brief description of the methods, results, and conclusions
Introduction	Structure is broad to specific: Introduce the main topic on a general level, state the question or unresolved problem, and then describe the objectives of the current study
	Provide background information by citing published sources.
Materials and Methods	Contains all of the relevant details to enable another trained scientist to repeat the procedure
	Does not contain routine procedures
Results	Contains text and visuals
	Text describes the results shown in each visual. Do not explain or interpret the results.
	Visuals include graphs, tables, photos, gel images, and so on, which contain the numerical or descriptive data.
Discussion	Structure is specific to broad: Interpret or explain the results, relate the results to published work, and describe the implications of the results in broad terms
References	List the full reference for each source cited.
	Do not list references that have not been cited.

Most scientific papers are divided into **standard sections**: title, authors' name(s), abstract, Introduction, Materials & Methods, Results, Discussion, and references (known as IMRD format). Readers of scientific papers like this format because they know where to look to find certain kinds of information. Confirm that these sections are in the **right order** in your paper. Then check that each section has the **appropriate content** (Table 5.1) by underlining each component on the printed pages or highlighting them on the computer screen (Hofmann 2010).

Do the math at least twice

Double-check your calculations and spreadsheet data entries. A mistake at this stage will have a negative domino effect, resulting in inaccurate figures or tables and a faulty discussion and interpretation of the results.

Organize each section

The phrases you underlined when you checked for content should provide a rough outline of each section. Do the **most important topics stand out**? Does the **order of the topics make sense** chronologically or sequentially? Is the order what your audience expects (for example, are the topics arranged from broad to specific in the Introduction section and specific to broad in the Discussion section)? Rearrange paragraphs so that the important topics can be identified easily, in an order that makes sense.

Make coherent paragraphs

Each paragraph should focus on only one topic. Make the topic sentence the first sentence in the paragraph. Follow the topic sentence with supporting sentences. Prepare readers for the topic in the next paragraph by linking the last sentence to the first sentence of the next paragraph. Alternatively, link the first sentence back to the previous paragraph.

Write meaningful sentences

Each sentence should **say something meaningful** and not repeat what was said before (**avoid redundancy**). Consider the following example.

> EXAMPLE: ~~From the data that has been gathered, a graph depicting the effect of various pH environments on the rate of catalase activity is represented. The graph displays the pH tested and the reaction rate. The data plotted is an accumulation of data from several lab sections. Through analyzing the graph I can see that~~ T~~t~~here ~~is~~ was no activity below a pH of 4 or above 10. Maximum catalase activity occurred at pH 7.

> EXPLANATION: The first three sentences do not convey anything substantive about the *results*. Only the last two sentences contain meaningful information.

To edit redundant sentences, take the best parts of those sentences and combine them into one concise sentence. Put yourself in your reader's shoes: Would you rather waste precious minutes wading through verbiage or get needed information with minimal effort?

Technical accuracy. Sentences that provide background information on a topic (as in the Introduction section), describe procedures (in the Materials and Methods section), or explain results (in the Discussion section) should be based on scientific fact. When in doubt, check your references, including secondary sources such as your textbook. Furthermore, make sure that your description of the results shown in each visual is accurate. In particular, pay attention to words like *increase* and *decrease*. Check that you did not mix up the results when you describe multiple data sets plotted on one graph.

Sentence length. Short sentences that contain only one idea are easy to understand. A text that contains nothing but short sentences, however, may be perceived as childish at best or hard to follow at worst. On the other hand, long, needlessly complex sentences obscure the main idea and slow comprehension. Aim for a mixture of short and long sentences in your writing. Use more words to explain a complex idea, but keep each sentence focused on just one idea.

Here are some examples of **needlessly complex sentences**.

FAULTY 1: *There are* two protein assays *that* are often used in research laboratories.

REVISION: Two protein assays are often used in research laboratories. (Avoid expletives.)

FAULTY 2: *It is interesting to note that* some enzymes are stable at temperatures above 60°C.

REVISION: Some enzymes are stable at temperatures above 60°C. (Avoid unnecessary introductory phrases.)

FAULTY 3: *The analyses were done on* the recombinant DNA to determine which piece of foreign DNA was inserted into the vector.

REVISION: The recombinant DNA was analyzed to determine which piece of foreign DNA was inserted into the vector. (Make *DNA*, not the *analyses*, the subject of the sentence.)

FAULTY 4: *We make the recommendation* that micropipettors be used to measure volumes less than 1 mL.

REVISION: We recommend that micropipettors be used to measure volumes less than 1 mL. (Replace sluggish noun phrases [nominalizations] with verb phrases.)

FAULTY 5:   These assays alone cannot *tell* what the protein concentration of a substance is.

REVISION: These assays alone cannot determine the protein concentration of a substance. (Replace colloquial expressions with precise alternatives.)

Emphasize the subject. Putting the subject at the beginning of the sentence makes the subject stand out. Position the verb close by so that there is no doubt about the subject's action.

Active and passive voice. In **active voice**, the subject *performs* the action. In **passive voice**, the subject *receives* the action. Consider the following example:

PASSIVE: The clam was opened by the sea star. (Emphasis on *clam*)

ACTIVE: The sea star opened the clam. (Emphasis on *sea star*)

Although the meaning is the same in both sentences, notice the difference in emphasis. In active voice, the emphasis is on the performer, and the action takes place in the direction the reader reads the sentence. Active voice is recommended by most style guides for reasons that include the following:

- It sounds more natural and is easier for the reader to process.
- It is shorter and more dynamic.
- There is no ambiguity about who/what the subject of the sentence is, or about who did the action.

Consider the following example:

PASSIVE: It was concluded from this observation that…

ACTIVE: I concluded from this observation that…

Passive voice leaves the reader wondering who drew the conclusion; active voice conveys this information clearly.

While active voice is generally preferred, passive voice may be more appropriate when *what* is being done is more important than *who* is doing it. For example:

> PASSIVE: Catalase was extracted from a potato. (Emphasis on *catalase*)

> ACTIVE: I extracted catalase from a potato. (Emphasis on *I*)

Notice the difference in emphasis. Is it really important to the success of the procedure that *you* did it, or does the emphasis belong on the catalase?

A paper that contains a mixture of active and passive voice is pleasant to read. Your decision to use active or passive voice in a sentence should ultimately be determined by clarity and brevity. In other words, use active voice to emphasize the subject and the fact that the subject is performing the action. Use passive voice when the action is more important than who is doing it.

Present or past tense. In scientific papers, present tense is used mainly in the following situations:

- To make generally accepted statements (for example, "Photosynthesis *is* the process whereby green plants produce sugars").
- When referring directly to a table or figure in your paper (for example, "Figure 1 *is* a schematic diagram of the apparatus").
- When stating the findings of published authors (for example, "Catalase HPII from *E. coli is* highly resistant to denaturation [Switala and others 1999]").

Past tense is used mainly in the following situations:

- To report your own work, especially in the abstract, Materials and Methods, and Results sections, because it remains to be seen if the scientific community accepts your work as fact (for example, "At temperatures above 37°C, catalase activity *decreased*").
- To cite another author's findings directly (for example, "Miller and others (1998) *found* that...").

Sentence order. Check that the order of the sentences makes sense chronologically or sequentially.

Continuity. A smooth transition from one sentence to the next is essential for reader comprehension. Consider the following example:

FAULTY: Catalase is an enzyme that breaks down hydrogen peroxide in both plant and animal cells. Low or high temperature can lower the rate at which the catalase can react with the hydrogen peroxide. In optimal conditions, the enzyme functions at a rate that will prevent any substantial buildup of the toxin. If the temperature is too low, the rate will be too slow, but high temperatures lead to the denaturation of the enzyme.

Where is the writer going with this paragraph? The sentences do not seem to flow because there is no guidance from the writer on how one sentence is related to the next. To improve flow, use connecting words such as *however*, *thus*, *although*, *in contrast*, *similarly*, *on the other hand*, *in addition to*, and *furthermore*. Another technique is to repeat a key word from the previous sentence to link one sentence to the next. Both of these strategies were used to revise the passage.

REVISION: Catalase is an enzyme that breaks down hydrogen peroxide in both plant and animal cells. One of the factors that affects the rate *of this reaction* is temperature. At optimal *temperatures*, the rate is sufficient to prevent substantial buildup of the toxic hydrogen peroxide. If the temperature is too low, *however*, the rate will be too slow, and hydrogen peroxide *accumulates* in the cell. *On the other hand*, high temperatures may denature the enzyme.

Choose your words carefully

Words are the basic organizational unit of language. The words you choose and how you arrange them in a sentence will determine how well you convey your meaning to your readers. Beware of the following word-level problems:

Keep related words together. Consider the following sentence taken from an English-language newspaper in Japan: "A committee was formed to examine brain death in the Prime Minister's office." Although brain death in the Prime Minister's office may be a political reality, what was really intended was, "A committee was formed in the Prime Minister's office to examine brain death."

Redundancy. **Redundancy** means using two or more words that mean the same thing. This problem is easily corrected by eliminating one of the redundant words (Table 5.2). Along with empty phrases, redundancy is a source of **wordiness**, using too many words to convey an idea.

Some people think that using more words makes them sound important. In science, however, wordiness should be avoided at all costs, because it indicates that the writer can't communicate clearly. For student writers whose papers are evaluated by instructors, lack of clarity translates into a low grade. Researchers and faculty members, whose reputation depends on the number and quality of their publications, simply cannot afford *not* to write clearly, because poorly written papers may be equated with shoddy scientific methods.

Empty phrases. Replace empty phrases with a concise alternative (Table 5.3). Put yourself in your reader's shoes. Which of the following two sentences (inspired by VanAlstyne 2005) would you rather read?

FAULTY: It is absolutely essential that you use a minimum number of words in view of the fact that your reader has numerous other tasks to complete at the present time.

REVISION: Write concisely, because your reader is busy.

Initially it is difficult to write in (and read) the terse, get-to-the-point style that characterizes scientific papers. With practice, however, you may come to appreciate this style because in a well-written paper, not a word is wasted. The benefit to you as a reader is that you extract a maximum amount of information from a minimum amount of text.

TABLE 5.2 Examples of redundancy	
Redundant	**Revised**
It is absolutely essential…	It is essential…
mutual cooperation	cooperation
basic fundamental concept	basic concept or fundamental concept
totally unique	unique
The solution was obtained and transferred…	The solution was transferred…

Ambiguous use of *this*, *that*, and *which*. Ambiguity results when *this*, *that*, or *which* could refer to more than one subject.

FAULTY: The data show that the longer the enzyme was exposed to the salt solution, the lower the enzyme activity in the assay. *This* means that the salt changes the conformation of the enzyme, *which* makes it less reactive with the substrate.

EXPLANATION: The subject of *this* and *which* is unclear.

REVISION: The longer the enzyme was exposed to the salt solution, the lower the enzyme activity in the

TABLE 5.3 Examples of empty phrases	
Empty	**Concise**
a downward trend	a decrease
a great deal of	much higher
a majority of	most
accounted for the fact that	because
as a result	so, therefore
as a result of	because
as soon as	when
at which time	when
at all times	always
at a much greater rate than	faster
at the present time, at this time	now, currently
based on the fact that	because
brief in duration	short, quick
by means of	by
came to the conclusion	concluded
despite the fact that, in spite of the fact that	although, though
due to the fact that, in view of the fact that	because
for this reason	so
in fact	*omit this phrase*

assay. Exposure to the salt solution may change the conformation of the enzyme, resulting in lower enzyme-substrate activity.

Ambiguous use of pronouns (*him, her, it, he, she, its*). Ambiguity results when a pronoun could refer to two possible antecedents.

FAULTY: With time, salt changes the conformation of the enzyme, which makes *it* less reactive with the substrate.

EXPLANATION: *It* could refer to *salt* or *enzyme*. To eliminate the ambiguity, replace *it* with the appropriate noun phrase.

TABLE 5.3 *Continued*	
Empty	**Concise**
functions to, serves to	*omit this phrase*
degree of	higher, more
in a manner similar to	like
in the amount of	of
in the vicinity of	near, around
is dependent upon	depends on
is situated in	is in
it is interesting to note that, it is worth pointing out that	*omit these kinds of unnecessary introductions*
it is recommended	I (we) recommend
on account of	because, due to
prior to	before
provided that	if
referred to as	called
so as to	to
through the use of	by, with
with regard to	on, about
with the exception of	except
with the result that	so that

REVISION: With time, salt changes the conformation of the enzyme, so that the enzyme can no longer react with its substrate.

Word usage. When you use the right words in the right situations, readers have confidence in your work. Use a standard dictionary whenever you are not sure about word usage. Consult your textbook and laboratory exercise for proper spelling and usage of technical terms. The following word pairs are frequently confused in students' lab reports.

absorbance, absorbency, observance *Absorbance* is how much light a solution absorbs; absorbance is measured with a spectrophotometer. *According to Beer's law, absorbance is proportional to concentration. Absorbency* is how much moisture a diaper or paper towel can hold. *Brand A paper towels show greater absorbency than Brand B paper towels. Observance* is the act of observing. *Government offices are closed today in observance of Independence Day.*

affect, effect *Affect* is a verb that means "to influence." Affect is never used as a noun. *Temperature affects enzyme activity. Effect* can be used either as a noun or a verb. When used as a noun, *effect* means "result." *We studied the effect of temperature on enzyme activity.* When used as a verb, effect means "to cause." *High temperature effected a change in enzyme conformation, which destroyed enzyme activity.*

alga, algae See plurals.

amount, number Use *amount* when the quantity cannot be counted. *The reaction rate depends on the amount of enzyme in the solution.* Use *number* if you can count individual pieces. *The reaction rate depends on the number of enzyme molecules in the solution.*

analysis, analyses See plurals.

bacterium, bacteria See plurals.

bind, bond *Bind* is a verb meaning "to link." *The active site is the region of an enzyme where a substrate binds. Bond* is a noun that refers to the chemical linkage between atoms. *Proteins consist of amino acids joined by peptide bonds. Bond* used as a verb means "to stick together." *This 5-minute epoxy glue can be used to bond hard plastic.*

complementary, complimentary   *Complementary* means "something needed to complete" or "matching." *The DNA double helix consists of complementary base pairs: A always pairs with T, and G with C. Complimentary* means "given free as a courtesy." *The brochures at the visitor's center are complimentary.*

confirmation, conformation   *Confirmation* means "verification." *I received confirmation from the postal service that my package had arrived. Conformation* is the three-dimensional structure of a macromolecule. *Noncovalent bonds help maintain a protein's stable conformation.*

continual, continuous   *Continual* means "going on repeatedly and frequently over a period of time." *The continual chatter of a group of inconsiderate students during the lecture annoyed me. Continuous* means "going on without interruption over a period of time." *The bacteria were grown in L-broth continuously for 48 hr.*

create, prepare, produce   *Create* is to cause to come into existence. *The artist used wood and plastic to create this sculpture. Prepare* means "to make ready." *The protein standards were prepared from a 50 mg/mL stock solution. Produce* means to make or manufacture. *The reaction between hydrogen peroxide and catalase produces water and oxygen.*

datum, data   See plurals.

different, differing   *Different* is an adjective that means "not alike." An adjective modifies a noun. *Different concentrations of bovine serum albumin were prepared. Differing* is the intransitive tense of "to differ," a verb that means "to vary." It is incorrect to replace the word *different* with *differing* in the preceding example, because *differing* implies that a single concentration changes depending on time or circumstance. This situation is highly unlikely with bovine serum albumin, which is quite stable under laboratory conditions! An acceptable use of *differing* is shown in the following example. *Bovine serum albumin solutions, differing in their protein content, were prepared.*

effect, affect   See affect, effect.

fewer, less   Use *fewer* when the quantity can be counted. *The reaction rate was lower, because there were fewer collisions between enzyme and substrate molecules.* Use *less* when the quantity cannot be counted. *The weight of this sample was less than I expected.*

formula, formulas, formulae   See plurals.

hypothesis, hypotheses   See plurals.

its, it's   *Its* is a possessive pronoun meaning "belonging to it." *The Bradford assay is preferred because of its greater sensitivity. It's* is a contraction of "it is." *The Bradford assay is preferred because it's more sensitive.* (Note: Contractions should not be used in formal writing.)

less, fewer   See fewer, less.

lose, loose   *Lose* means to misplace or fail to maintain something. *An enzyme may lose its effectiveness at high temperatures. Loose* means "not tight." *When you autoclave solutions, make sure the lid on the bottle is loose.*

lowered, raised   Both of these are transitive verbs, which means that they require a direct object (a noun to act on). **Wrong**: *The fish's body temperature lowered in response to the cold water.* **Right**: *The cold water lowered the fish's body temperature.*

media, medium   See plurals.

observance   See absorbance, absorbency, observance.

phenomenon, phenomena   See plurals.

plurals   The plural and singular forms of some words used in biology are given in Table 5.4. A common mistake with these words is not making the subject and verb agree. Some disciplines treat *data* as singular, but scientists subscribe to the strict interpretation that *data* is plural. The data *show…* (not *shows*) is correct.

prepare   See create, prepare, produce.

produce   See create, prepare, produce.

raised, lowered   See lowered, raised.

TABLE 5.4 Singular and plural of words frequently encountered in biology

Singular	Plural
alga	algae
analysis	analyses
bacterium	bacteria
criterion	criteria
datum (rarely used)	data
formula	formulas, formulae
hypothesis	hypotheses
index	indexes, indices
medium	media
phenomenon	phenomena
ratio	ratios

ratio, ration *Ratio* is a proportion or quotient. *The ratio of protein in the final dilution was 1:5. Ration* is a fixed portion, often referring to food. *The Red Cross distributed rations to the refugees.*

strain, strand A *strain* is a line of individuals of a certain species, usually distinguished by some special characteristic. *The lacI⁻ strain of E. coli produces a nonfunctional repressor protein.* A *strand* is a ropelike length of something. *The strands of DNA are held together with hydrogen bonds.*

than, then *Than* is an expression used to compare two things. *Collisions between molecules occur more frequently at high temperatures than at low temperatures. Then* means "next in time." *First 1 mL of protein sample was added to the test tube. Then 4 mL of biuret reagent was added.*

that, which Use *that* with restrictive clauses. A restrictive clause limits the reference to a certain group. Use *which* with nonrestrictive clauses. A nonrestrictive clause does not limit the reference, but rather provides additional information. Commas are used to set off nonrestrictive clauses but not restrictive clauses. Consider the following examples:

EXAMPLE 1:   The Bradford assay, which is one method for measuring protein concentration, requires only a small amount of sample. (*Which* begins a phrase that provides additional information, but is not essential to make a complete sentence.)

EXAMPLE 2:   Enzyme activity decreased significantly, which suggests that the enzyme was denatured at 50°C. (*Which* refers to the entire phrase *Enzyme activity decreased significantly,* not to any specific element.)

EXAMPLE 3:   The samples that had high absorbance readings were diluted. (*That* refers specifically to *the samples.*)

various, varying   *Various* is an adjective that means "different." *Various hypotheses were proposed to explain the observations. Varying* is a verb that means "changing." *Varying the substrate concentration while keeping the enzyme concentration constant allows you to determine the effect of substrate concentration on enzyme activity.* Analogous to *different, differing,* replacing the word *various* with *varying* in the preceding example changes the meaning of the sentence. *Varying* implies that a single hypothesis changes depending on time or circumstance. *Various* implies that different hypotheses were proposed.

Jargon and scientific terminology. **Jargon** refers to words and abbreviations used by specialists. Whenever you use terms that may be unfamiliar to your audience, define them. Always write out the full expression when first using an abbreviation. Scientific words that you learned in class are *not* jargon and should be used in your writing. When you use scientific terminology correctly, your readers have confidence in your knowledge.

Clichés and slang. **Clichés** are tired, worn out expressions that have no place in an exciting field like biology. Slang should not be used, because colloquial language is not appropriate in formal, written assignments like lab reports.

Gender-neutral language. Years ago, it was customary, for the sake of simplicity, to use masculine pronouns to refer to antecedents that could be masculine or feminine, but that use of language is no longer accepted.

SEXIST: The clarity with which a biology student writes *his* lab reports affects *his* grade.

This practice is no longer considered to be politically correct. One solution that preserves equality, but makes sentences unnecessarily complex, is to include both masculine and feminine pronouns, as in the following example.

EQUAL BUT   The clarity with which a biology student writes
AWKWARD: *his or her* lab reports affects *his or her* grade.

Two better alternatives are to make the antecedent plural (revision 1) or to rewrite the sentence to avoid the gender issue altogether (revision 2).

REVISION **1:**   The clarity with which biology students write *their* lab reports affects *their* grade.

REVISION **2:**   *Writing clearly* has a positive effect on a biology student's grade. (Change the subject from *biology student* to *writing clearly*.)

Construct memorable visuals

Visuals often make the difference in how well you convey your meaning to your readers or listeners. Make sure you use the **appropriate visual** for the data (see "Organizing your data" on pp. 72–75). Make sure **every visual serves a purpose**, because unnecessary visuals only dilute the significance of your message. Check that the **visuals are positioned in the right order** and that **each visual is described** in the text.

Proofreading: The Home Stretch

Proofreading is the last stage of revision. Like editing, it requires intense focus and slow, careful reading to find mistakes in format, spelling, punctuation, and grammar. **Grammar** refers to the rules that deal with the form and structure of words and their arrangement in sentences. See Hacker and Sommers (2012) or Lunsford (2010a, 2010b, 2011) for a more comprehensive treatment of the subject.

Make subjects and verbs agree

We learn early on in our formal education to make the verb agree with the subject. Most of us know that *the sample was…*, but that *the samples were….* Most errors with subject–verb agreement occur when there are words *between* the subject and the verb, as in the following example.

> EXAMPLE: The *height* of the lamps *was* (not *were*) adjusted to make the temperature inside the dome about 28°C.

When you write complex sentences, ask yourself what the subject of the sentence is. Look for the verb that goes with that subject. Then, mentally remove the words in between the two, and make the subject and its verb agree.

A second situation in which subject–verb agreement becomes confusing is when there are two subjects joined by *and,* as in the following example.

> EXAMPLE: An enzyme's amino acid *sequence and* its three-dimensional *structure make* (not *makes*) the enzyme–substrate relationship unique.

Compound subjects joined by *and* are almost always plural.

A third situation involving subject–verb agreement is that when numbers are used in conjunction with units, the *quantity* is considered to be *singular,* not plural.

> EXAMPLE: To extract the enzyme, 50 g of potatoes *was* (not *were*) homogenized with 50 mL of cold, distilled water.

Write in complete sentences

A complete sentence consists of a subject and a verb. If the sentence starts with a subordinate word or words such as *after, although, because, before, but, if, so that, that, though, unless, until, when, where, who,* or *which,* however, another clause must complete the sentence.

> FAULTY 1: High temperatures destroy the three-dimensional structure of enzymes. Thus changing the effectiveness of the enzymes. (The second "sentence" is a fragment.)

> REVISION: High temperatures destroy the three-dimensional structure of enzymes, thus changing their effectiveness. (Combine the fragment with the previous sentence, changing punctuation as needed.)

> FAULTY 2: The standard curve for the biuret assay was used to determine the protein concentration of the serial dilutions of the egg white. Although only

those dilutions whose protein concentrations fell within the sensitivity range of the assay were multiplied by the dilution factor to give the original concentration of the egg white. (The second "sentence" is a fragment.)

REVISION: The standard curve for the biuret assay was used to determine the protein concentration of the serial dilutions of the egg white. Only those dilutions whose protein concentrations fell within the sensitivity range of the assay were multiplied by the dilution factor to give the original concentration of the egg white. (Delete the subordinate word[s] to make a complete sentence.)

Revise run-on sentences

Run-on sentences consist of two or more independent clauses joined without proper punctuation. Each independent clause could stand alone as a complete sentence. Run-on sentences are common in first drafts, where your main objective is to get your ideas down on paper (or electronic media, if you use a computer). When you revise your first draft, however, use one of the following strategies to revise run-on sentences:

- Insert a comma and a coordinating conjunction (*and, but, or, nor, for, so,* or *yet*).
- Use a semicolon or possibly a colon.
- Make two separate sentences.
- Rewrite the sentence.

FAULTY 1:   The class data for the Bradford method were scattered, those for the biuret method were closer.

REVISION A: The class data for the Bradford method were scattered, but those for the biuret method were closer. (Use a coordinating conjunction.)

REVISION B: The class data for the Bradford method were scattered; those for the biuret method were closer. (Use a semicolon.)

FAULTY 2:   The readings from the spectrophotometer should show a correlation between protein concentration and absorbance, this is Beer's law, which relates absorbance to the path length of light along with molar concentration of a solute and the molar coefficient. (Fused sentence)

REVISION A: The readings from the spectrophotometer should show a correlation between protein concentration and absorbance; this is Beer's law, which relates absorbance to the path length of light along with molar concentration of a solute and the molar coefficient. (Use a semicolon to separate the two clauses.)

REVISION B: The readings from the spectrophotometer should show a correlation between protein concentration and absorbance. This relationship is described by Beer's law, which relates absorbance to the path length of light along with molar concentration of a solute and the molar coefficient. (Make two separate sentences.)

FAULTY 3:   An increase in enzyme concentration increased the reaction rate as did an increase in substrate concentration, so the concentrations of the molecules have an influence on how the enzyme reacts.

REVISION A: As enzyme concentration and substrate concentration increased, so did the reaction rate. (Rewrite the sentence. The second half of the original sentence was deleted because it says nothing meaningful.)

REVISION B: Enzyme and substrate concentration influence enzyme reaction rate: an increase in enzyme or substrate concentration increased reaction rate. (Use a colon.)

Spelling

Spell checkers in word processing programs are so easy to use that there is really no excuse for *not* using them. Just remember that spell checkers may not know scientific terminology, so consult your textbook or laboratory manual for correct spelling. In some cases, the spell checker may even try to get you to change a properly used scientific word to an inappropriate word that happens to be in its database (for example, *absorbance* to *absorbency*).

The following poem is an example of how indiscriminate use of the spell checker can produce garbage:

Wrest a Spell

Eye halve a spelling chequer
It came with my pea sea
It plainly marques four my revue
Miss steaks eye kin knot sea.

Eye strike a key and type a word
And weight four it two say
Weather eye am wrong oar write
It shows me strait a weigh.

As soon as a mist ache is maid
It nose bee fore two long
And eye can put the error rite
Its rare lea ever wrong.

Eye have run this poem threw it
I am shore your pleased two no
Its letter perfect awl the weigh
My chequer tolled me sew.

— Sauce unknown

Spell checkers will also not catch mistakes of usage, for example *form* if you really meant *from*. Print out your document and proofread the hard copy carefully.

Punctuation

The purpose of punctuation marks is to divide sentences and parts of sentences to make the meaning clear. A few of the most common punctuation marks and their uses are described in this section. For a more comprehensive, but still concise, treatment of punctuation, see Hacker and Sommers (2012), or Lunsford (2010a, 2010b, 2011).

The comma. The comma inserts a pause in the sentence in order to avoid confusion. Note the ambiguity in the following sentence:

> While the sample was incubating the students prepared the solutions for the experiment.

A comma *should* be used in the following situations:

1. To connect two independent clauses that are joined by *and, but, or, nor, for, so,* or *yet.* An independent clause contains a subject and a verb, and can stand alone as a sentence.

> EXAMPLE: Feel free to call me at home, but don't call after 9 p.m.

2. After an introductory clause, to separate the clause from the main body of the sentence.

> EXAMPLE: Although she spent many hours writing her lab report, she earned a low grade because she forgot to answer the questions in the laboratory exercise.

A comma is not needed if the clause is short.

> EXAMPLE: Suddenly the power went out.

3. Between items in a series, including the last two.

> EXAMPLE: Enzyme activity is affected by factors such as substrate concentration, pH, temperature, and salt.

4. Between coordinate adjectives (if the adjectives can be connected with *and*).

> EXAMPLE: The students' original, humorous remarks made my class today particularly enjoyable. (*Original and humorous remarks* makes sense.)

A comma is not needed if the adjectives are cumulative (if the adjectives cannot be connected with *and*).

> EXAMPLE: The three tall muscular students look like football players. (It would sound strange to say *three and tall and muscular students.*)

5. With *which,* but not *that* (see Word usage: that, which, pp. 109–110)

6. To set off conjunctive adverbs such as *however, therefore, moreover, consequently, instead, likewise, nevertheless, similarly, subsequently, accordingly,* and *finally.*

> EXAMPLE: Instructors expect students to hand in their work on time; however, illness and personal emergencies are acceptable excuses.

7. To set off transitional expressions such as *for example, as a result, in conclusion, in other words, on the contrary,* and *on the other hand.*

> EXAMPLE: Chuck participates in many extracurricular activities in college. As a result, he rarely gets enough sleep.

8. To set off parenthetical expressions. Parenthetical expressions are statements that provide additional information; however, they interrupt the flow of the sentence.

> EXAMPLE: Fluency in a foreign language, as we all know, requires years of instruction and practice.

A comma *should not* be used in the following situations.

1. After *that,* when *that* is used in an introductory clause

> EXAMPLE: The student could not believe that he lost points on his laboratory report because of a few spelling mistakes.

2. Between cumulative adjectives, which are adjectives that would not make sense if separated by the word *and* (see Item 4 in preceding list)

The semicolon. The semicolon inserts a stop between two independent clauses not joined by a coordinating conjunction (*and, but, or, nor, for, so,* or *yet*). Each independent clause (one that contains a subject and a verb) could stand alone as a sentence, but the semicolon indicates a closer relationship between the clauses than if they were written as separate sentences.

> EXAMPLE: Outstanding student-athletes use their time wisely; this trait makes them highly sought after by employers.

A semicolon is also used to separate items in a series in which the items are already separated by commas.

> EXAMPLE: Participating in sports has many advantages.
> First, you are doing something good for your
> health; second, you enjoy the camaraderie of
> people with a common interest; third, you learn
> discipline, which helps you make effective use of
> your time.

The colon. The colon is used to call attention to the words that follow it. Some conventional uses of a colon are shown in the following examples.

> Dear Sir or Madam:
> 5:30 P.M.
> 2:1 (ratio)

In references, to separate the place of publication and the publisher, as in

> Sunderland (MA): Sinauer Associates, Inc.

A colon is often used to set off a list, as in the following example.

> EXAMPLE: Catalase activity has been found in the following
> vegetables: potatoes, leeks, parsnips, onions, zuc-
> chini, carrots, and broccoli.

A colon *should not* be used when the list follows the words *are, consist of, such as, including,* or *for example.*

> EXAMPLE: Catalase activity has been found in vegetables
> such as potatoes, leeks, parsnips, onions,
> zucchini, carrots, and broccoli.

The period. The period is used to end all sentences except questions and exclamations. It is also used in American English for some abbreviations, for example, *Mr., Ms., Dr., Ph.D., i.e.,* and *e.g.*

Parentheses. Parentheses are used mainly in two situations in scientific writing: to enclose supplemental material and to enclose references to visuals or sources. Use parentheses sparingly because they interrupt the flow of the sentence.

> EXAMPLE: Human error (failure to make the solutions cor-
> rectly, arithmetic errors, and failure to zero the
> spectrophotometer) was the main reason for the
> unexpected results.

REFERENCE TO VISUAL:	There was no catalase activity above 70°C (Figure 1).

CITATION-SEQUENCE SYSTEM:	C-fern spores do not germinate in the dark (1).

NAME-YEAR SYSTEM:	C-fern spores do not germinate in the dark (Cooke and others 1987).

The dash. The dash is used to set off material that requires special emphasis. To make a dash on the computer, type two hyphens without a space before, after, or in between. In some word processing programs, the two hyphens are automatically converted to a solid dash.

Similar to commas and parentheses, a pair of dashes may be used to set off supplemental material.

> EXAMPLE: Human error--failure to make the solutions correctly, arithmetic errors, and failure to zero the spectrophotometer--was the main reason for the unexpected results. (If the word processing program has been set up to convert the two hyphens to a solid dash, the sentence looks like this: Human error—failure to make…spectrophotometer—was the main reason…)

Similar to a colon, a single dash calls attention to the information that follows it.

> EXAMPLE: Catalase activity has been found in many vegetables—potatoes, leeks, parsnips, onions, zucchini, carrots, and broccoli.

If an abrupt or dramatic interruption is desired, use a dash. If the writing is more formal or the interruption should be less conspicuous, use one of the other three punctuation marks. However, do not replace a pair of dashes with commas when the material to be set off already contains commas, as in the following example.

> EXAMPLE: The instruments that she plays—oboe, guitar, and piano—are not traditionally used in the marching band.

Abbreviations

The CSE Manual (2006) defines standard abbreviations for authors and publishers in the sciences and mathematics. Some of the terms and abbreviations that you may encounter in introductory biology courses are shown in Table 5.5. Take note of spacing, case (capital or lowercase letters), and punctuation use. Except where noted, the symbols are the same for singular and plural terms (for example, 30 min *not* 30 mins).

Numbers

Numbers are used for quantitative measurements. In the past, numbers less than 10 were spelled out, and larger numbers were written as numerals. The modern scientific number style recommended in the CSE Manual (2006) aims for a more consistent usage of numbers. The rules are as follows:

1. Use numerals to express any *quantity*. This form increases their visibility in scientific writing, and emphasizes their importance.

   - Cardinal numbers, for example, 3 observations, 5 samples, 2 times
   - In conjunction with a unit, for example, 5 g, 0.5 mm, 37°C, 50%, 1 hr. Pay attention to spacing, capitalization, and punctuation (see Table 5.5).
   - Mathematical relationships, for example, 1:5 dilution, 1000× magnification, 10-fold

2. Spell out numbers in the following cases:

   - When the number begins a sentence, for example, *"Fifty g of potatoes was* (not *were*) *homogenized."* rather than *"50 g of potatoes was homogenized."* Alternatively, restructure the sentence so that the number does not begin the sentence. Notice that when numbers are used in conjunction with units, the quantity is considered to be singular, not plural.
   - When there are two adjacent numbers, retain the numeral that goes with the unit, and spell out the other one. An example of this is *The solution was divided into four 250-mL flasks.*
   - When the number is used in a nonquantitative sense, for example, *one of the treatments, the expression approaches zero, one is required to consider…*
   - When the number is an ordinal number less than 10, and when the number expresses rank rather than quantity, for example, *the second time, was first discovered.*

	Symbol or	
Term	Abbreviation	Example

TABLE 5.5 Standard abbreviations in scientific writing

Term	Symbol or Abbreviation	Example
Latin words and phrases [The CSE Manual (2006) recommends that Latin-words be replaced with English equivalents.]		[The Latin word may be replaced with the English equivalent given in brackets.]
circa (approximately)	ca.	The lake is ca. [approx.] 300 m deep.
et alii (and others)	*et al.*	Jones *et al.* [and others] (1999) found that …
et cetera (and so forth)	etc.	pH, alkalinity, etc. [and other characteristics] were measured.
exempli gratia (for example)	e.g.	Water quality characteristics (e.g., [for example,] pH, alkalinity) were measured.
id est (that is)	i.e.	The enzyme was denatured at high temperatures, i.e., the enzyme activity was zero. [Because the enzyme was denatured at high temperatures, the enzyme activity was zero.]
nota bene (take notice)	NB	NB [Important!]: Never add water to acid when making a solution.
Length		
nanometer (10^{-9} meter)	nm	*Note:* There is a space between the number and the abbreviation. There is no period after the abbreviation.
micron (10^{-6} meter)	μm	
millimeter (10^{-3} meter)	mm	
centimeter (10^{-2} meter)	cm	
meter	m	450 nm, 10 μm, 2.5 cm
Mass		
nanogram (10^{-9} gram)	ng	*Note*: There is a space between the number and the abbreviation. There is no period after the abbreviation.
microgram (10^{-6} gram)	μg	
milligram (10^{-3} gram)	mg	
gram	g	
kilogram (10^{3} gram)	kg	450 ng, 100 μg, 2.5 g, 10 kg

TABLE 5.5 *Continued*		
Term	**Symbol or Abbreviation**	**Example**
Volume		
microliter (10^{-6} liter)	μl or μL	*Note:* There is a space between the number and the abbreviation. There is no period after the abbreviation.
milliliter (10^{-3} liter)	ml or mL	
liter	l or L	
cubic centimeter (ca. 1 mL)	cm^3	450 μl or 450 μL, 0.45 ml or 0.45 mL, 2 l or 2 L
Time		
seconds	s or sec	*Note:* There is a space between the number and the abbreviation. There is no period after the abbreviation (unless the unit ends a sentence).
minutes	min	
hours	h or hr	
days	d	
		60 s or 60 sec, 60 min, 24 h or 24 hr, 1 d
Concentration		
molar (U.S. use)	M	TBS contains 0.01 M Tris-HCl, pH 7.4 and 0.15 M NaCl.
molar (SI units)	mol L^{-1}	
parts per thousand	ppt	Brine shrimp can be raised in 35 ppt seawater.
Other		
degree(s) Celsius	°C	15°C (no space between number and symbol)
degree(s) Fahrenheit	°F	59°F (no space between number and symbol)
diameter	diam.	pipe diam. was 10 cm
figure, figures	Fig., Figs.	As shown by Fig. 1, …
foot-candle	fc or ft-c	500 fc or 500 ft-c
maximum	max	The max enzyme activity was found at 36°C.
minimum	min	The min temperature of hatching was 12°C.
mole	mol	
percent	%	95% (no space between number and symbol)
species (sing.)	sp.	*Tetrahymena* sp.
species (plur.)	spp.	*Tetrahymena* spp.

TABLE 5.6 Checklist for proofreading format	
Category	**Check for**
Section headings	Correct order, consistent format, not separated from section body
Lists (bulleted or numbered)	Sequential numbering and consistent style, parallelism in sentence structure, consistent indentation for each level
Figures and tables	Sequential numbering in the order they are described
In-text references to figure and table numbers	Correspondence with the actual figures and tables
In-text citations	One-to-one correspondence with the end references, correct formatting
End references	One-to-one correspondence with the in-text citations, completeness, correct formatting
Headers and footers (if needed)	Correct position on each page
Page numbers	Sequential (check especially after section breaks in Microsoft Word)
Typography	Consistent typeface, font size, and spacing

- When the number is a fraction used in running text, for example, *one-half of the homogenate, nearly three-quarters of the plants.* When the precise value of a fraction is required, however, use the decimal form, for example, *0.5 L* rather than *one-half liter.*

3. Use scientific notation for very large or very small numbers. For the number 5,000,000, write 5×10^6, not *5 million*. For the number *0.000005*, write 5×10^{-6}.

4. For decimal numbers less than one, always mark the ones column with a zero. For example, write *0.05*, not *.05*.

Format

Most university writing centers and professional editors recommend proofreading your paper in multiple "passes," looking for one kind of mistake in each pass (The Writing Center at UNC Chapel Hill 2012; Cook Counseling Center – Virginia Tech 2009; Every 2012). This strategy works particularly well for finding formatting errors, which are much easier to detect on printed pages than on the computer screen (Table 5.6). Check for potential errors in the following areas:

- Section headings
- Bulleted or numbered lists
- Figures and tables, including their in-text references
- In-text citations
- End references
- Headers and footers
- Page numbers
- Typography

Get Feedback

When we are engrossed in our work, we may fail to recognize that what is obvious to us is not obvious to an "outsider." That is where feedback from someone who is familiar with the subject matter comes in handy. If your instructor allows it, ask your lab partner, another classmate, or your teaching assistant to review your paper. Return the favor by reviewing someone else's. You may also get valuable feedback from a writing expert at your school's writing center.

The questions your reviewer will focus on are as follows:

- Do I know what the writer is trying to accomplish with this paper? Is the purpose clear?
- What questions or concerns do I have about this paper? Are there sections that were difficult to follow? Are the organization, content, flow, and level appropriate for the intended audience?
- What suggestions can I offer the writer to help him/her clarify the intended meaning?
- What do I like about the paper? What are its strengths?

Tips for being a good peer reviewer

There are two issues with which you may struggle when you are asked to review your classmate's paper: (1) I'm not confident that I know the "right" answer or know enough about the writing process to give good suggestions, and (2) I don't want to hurt the writer's feelings. These are valid concerns, and resolving them will require, first, a willingness to learn as much about writing scientific papers as possible, and second, the attitude that if something is unclear to you, it may also be unclear to other readers. With each paper you review, you will gain more confidence in your ability to give constructive feedback. In the meantime, however, a good rule of thumb is to give the kinds of suggestions and consideration that you would like to receive on your own paper.

When reviewing electronic files, the **New Comment** and **Track Changes** commands on the **Review** tab in Microsoft Word are very useful (see Appendix 1, pp. 190–192). **New Comment** allows the reviewer to make a comment or query, without editing the text itself, off to the side of the main text. When **Track Changes** is turned on, the reviewer's suggested changes, typed right into the body of the text, appear in a different color, and any deletions are recorded to the side of the main text. The author can then accept or reject the changes. Think of the peer review process as a team sport: the reviewer is not challenging the writer's right to be on the team. The two are working together to get the best possible result.

Here are some concrete tips for being a good peer reviewer:

- Talk to the writer about his or her objectives, questions and concerns, parts that need specific feedback, and perceived strengths and weaknesses.
- Use the "Biology Lab Report Checklist" (pp. 127–128) for content.
- Look over the "Laboratory Report Mistakes" list (pp. 138–141) for common mistakes.
- Mark awkward sentences, spelling and punctuation mistakes, and formatting errors. Do not feel you have to rewrite individual sentences—that is the writer's job.
- Ask questions. Let the writer know where you can't follow his/her thinking, where you need more examples, where you expect more detailed analysis, and so on.
- Do not be embarrassed about making lots of comments; the author does not have to accept your suggestions. On the other hand, if you say only good things about the paper, how will the writer know whether the paper is accomplishing the desired objectives?

You can fine-tune your proofreading skills on any text. You may recognize some of your own problems in other people's writing, and, with persistence and practice, you will find creative solutions to correct these problems. **Keep a log of the problems that recur in your writing** and review them from time to time. Repetition builds awareness, which will help you achieve greater clarity in your writing.

Have an informal discussion with your peer reviewer

Sometimes the comments made by the peer reviewer are self-explanatory. Other times, however, the peer reviewer cannot respond to certain parts of the paper, because more information is required. Under these circumstances, an informal discussion between the writer and the reviewer is helpful. There are two important rules for this discussion:

- First, the writer talks and the reviewer listens. The objective is to help the writer express exactly what he/she wants to say in the paper.
- Second, the reviewer talks, in nonjudgmental terms, about which parts of the paper were readily understandable and which parts were confusing. The reviewer does not have to be an experienced writer to do this—no two people have exactly the same life experiences, and there is always something positive you can learn from looking at your writing from someone else's perspective.

Biology Lab Report Checklist

TITLE (p. 80)

☐   Descriptive and concise

AUTHORS

☐   Each author's first name is followed by his/her surname

ABSTRACT (pp. 79–80)

☐   Contains an introduction (background and objectives)
☐   Contains brief description of methods
☐   Contains results
☐   Contains conclusions

INTRODUCTION (pp. 78–79)

☐   Starts with a general introduction to the topic
☐   Contains a question or unresolved problem
☐   References support the background information
☐   Selected references are directly relevant to your study
☐   Citation format is correct
☐   Citations are paraphrased. Direct quotations are not used.
☐   Objectives of the study are clearly stated.

MATERIALS AND METHODS

☐   Contains all relevant information to enable the reader to repeat the procedure (pp. 52–53).
☐   Routine procedures are not explained (p. 52)
☐   Written in paragraph form (not a numbered list) and in complete sentences (pp. 51–55)
☐   Written in past tense (p. 51)
☐   Written in passive voice (active voice is allowed in some disciplines) (pp. 51–52)
☐   Materials are not listed separately (p. 53).
☐   No preview is given of how the data will be organized or interpreted (pp. 54–55)

RESULTS

- ☐ Figures and tables are present. Figure caption goes below figure; table caption goes above table (pp. 61–72).
- ☐ Text is present. Every sentence is meaningful (p. 58).
- ☐ Reference is made to each table and figure, and the results are described in words (p. 59).
- ☐ Figure and table titles are informative and can be understood apart from the text (pp. 61–72).
- ☐ No explanation is given for the results (p. 57).

DISCUSSION (pp. 76–78)

- ☐ Results are briefly restated.
- ☐ Results are explained and interpreted.
- ☐ Results are related to published work.
- ☐ Errors and inconsistencies are pointed out.
- ☐ Implications of the results are described.

REFERENCES (pp. 81–94)

- ☐ References consist mostly of primary journal articles, not textbooks or websites.
- ☐ Reference format is correct and complete.
- ☐ All end references have been cited in the text. All in-text citations have been included in the References section.

REVISION

- ☐ All questions from the laboratory exercise have been answered.
- ☐ Calculations and statistics have been double-checked (p. 97).
- ☐ Overall structure (pp. 96–97)
- ☐ Figures and tables (pp. 61–72 and 111)
- ☐ Sections, paragraphs, sentences, and words (pp. 98–111)
- ☐ Word usage (pp. 106–110)
- ☐ Grammar (pp. 111–114)
- ☐ Spelling (p. 115)
- ☐ Punctuation (pp. 115–119)
- ☐ Standard abbreviations (pp. 120–122)
- ☐ Numbers (pp. 120 and 123)
- ☐ Format (pp. 123–124)

Biology Lab Report Checklist

Chapter 6

A "GOOD" SAMPLE STUDENT LABORATORY REPORT

The laboratory report in this chapter was written by Lynne Waldman during her first year at Bucknell University, in an introductory course for biology majors. Lynne and her lab partners designed and carried out an original project in which they investigated the effect of a fungus on the growth of bean, pea, and corn plants.

Lynne's report has many of the characteristics of a well-written scientific paper. When you look over her presentation, notice the style and tone of her writing, as well as the format of the paper. The comments and annotations in the margins alert you to important points to keep in mind when you write your laboratory report.

The presentation here has been typeset to fit this book and to accommodate the annotations. Your report should be formatted to fit standard 8.5 × 11 inch paper. Unless you are instructed otherwise, use a serif type (Times Roman is standard), double space, and leave *at least* 1 inch of margin all around.

For details on how to format documents in Microsoft Word 2010, see "The Home Tab" (pp. 176–180) and "The Page Layout Tab" (p. 188) in Appendix 1.

Title is informative.

Write author's name first followed by lab partners' names.

Label sections of lab report clearly.

Provide background information.

State purpose of current experiment.

Do not cite sources in abstract.

Briefly describe methods.

Do not refer to any figures.

Describe results.

Briefly explain the results or state your conclusions.

Limit abstract to a maximum of 250 words.

Italicize Latin names.

Provide background information.

The Effects of the Fungus *Phytophthora infestans* on Bean, Pea, and Corn Plants

Lynne Waldman, Partner One, Partner Two

Abstract

Phytophthora infestans is a fast-spreading, parasitic fungus that caused the infamous potato blight by devastating Ireland's crops in the 1840s. *P. infestans* also causes late blight in tomato plants, a relative of the potato. In this experiment, the effects of *P. infestans* on *Phaseolus* variety long bush bean, *Zea mays* (corn), and *Pisum sativum* (pea) were studied. The soil surrounding the roots of 18-day old plants was injected with *P. infestans* cultured in an L-broth medium. Plant height, number of leaves, and leaf angle were measured for each plant during the next 8 days. Chlorophyll assays were performed prior to exposure, and on the eighth day after exposure to the fungus. The plants were also examined for black or brown leaf spots characteristic of late blight infections. The results showed that *P. infestans* had no apparent effect on the bean, corn, and pea plants. One reason for this may be that there were no fungus zoospores in the L-broth medium. More probably, however, *P. infestans* may be a species-specific pathogen that cannot infect bean, corn, or pea plants.

Introduction

Originating in Peruvian-Bolivian Andes, the potato (*Solanum tuberosum*) is one of the world's four most important food crops (along with wheat, rice, and corn). Cultivation of potatoes began in South America over 1,800 years ago, and through the Spanish conquistadors, the tuber was introduced into Europe in the second half of the 1600s. By the beginning of the 18th century, the potato was widely grown in Ireland, and the country's economy heavily relied on the potato crop. In the middle of the 19th century, Ireland's potato crop suffered widespread late blight disease caused by *Phytophthora infestans,* a species of pathogenic plant fungus. Failure of the potato crop because of late blight resulted in the Irish potato famine. The famine led to widespread starvation and the death of about a million Irish people.

The potato continues to be one of the world's main food crops. However, *P. infestans* has reemerged in a chemical-resistant form in the United States, Canada, Mexico, and Europe (McElreath, 1994). Late blight caused by the new strains is costing growers worldwide about $3 billion annually. The need to apply chemical fungicides eight to ten times a season further increases the cost to the grower (Stanley, 1994 and Stanley, 1997). *P. infestans* is thus an economically important pathogen.

P. infestans, which can destroy a potato crop in the field or in storage, thrives in warm, damp weather. The parasitic fungus causes black or purple lesions on a potato plant's stem and leaves. As a result of infection by this fungus, the plant is unable to photosynthesize, develops a slimy rot, and dies. *P. infestans* similarly infects the tomato plant (*Lycopersicon esculentum*) (Brave New Potato, 1994).

The purpose of the present experiment was to determine the effects of *P. infestans* on plant height, number of leaves, leaf angle, and chlorophyll content of three agriculturally important plants: *Phaseolus* variety long bush beans, *Zea mays* (corn), and *Pisum sativum* (peas). Symptoms of fungal infection were assumed to be similar to that in potatoes.

Materials and Methods

Phaseolus variety long bush bean, *Zea mays* (corn) and *Pisum sativum* (pea) seeds were soaked overnight in tap water. Fifteen randomly chosen seeds of each species were planted 1 cm beneath the surface in three separate trays containing 10 cm of potting soil. Another set of trays, which was to be the control group, was prepared in the same fashion. All

Use proper citation format (e.g., Name-Year system).

Do not use direct quotations. Paraphrase source text and cite the source in parentheses.

Use an abbreviated title when no author is given.

State purpose of experiment clearly.

Write Materials and Methods section in past tense.

Provide sufficient detail to allow the reader to repeat the experiment.

the experimental plants were placed in one fume hood, and all the control plants were placed in relative positions in another fume hood in the same room. The plants were exposed to the ambient light intensity in the hood (153 fc) and air current 24 hrs a day, and were watered lightly daily. The plants were allowed to germinate and grow for 18 days.

Phytophthora infestans on potato dextrose agar was obtained from Carolina Biological Supply House. At day 10 of the plant growth regime, pieces of agar on which the fungus was growing were transferred to L-broth. L-broth consisted of 5 g yeast extract, 10 g tryptone, 1 g dextrose, and 10 g NaCl dissolved in distilled water, and adjusted to pH 7.1, to make 1 L of medium. The medium was sterilized before adding the fungal culture. After 4 days in L-broth, 6 mL of the fungal culture was injected into the soil around the roots of each 18-day old plant. Six mL of L-broth without *P. infestans* was injected into the soil of the control plants. All plants were then allowed to grow for another 8 days.

Every other day after treatment with *P. infestans*, plant height and number of leaves were measured for both the control and the experimental plants. Plant height was measured from the soil to the apical meristem of the plant. Leaf angle (as shown in Figure 1) of the largest, lowest leaf on each plant was measured three times, once prior to injection, once 4 days after injection, and once 8 days after injection. Leaf angle was measured in order to deter-

Include figures in Materials and Methods section if they help clarify the methodology.

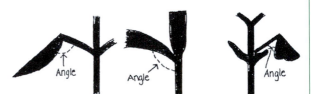

Figure 1 Leaf angle as measured in bean, corn, and pea plants

mine if *P. infestans* causes wilting in the three plant species. In addition, the plant was examined visually for the presence of any leaf spots.

Chlorophyll assays were performed on one plant from each tray prior to injection and on the eighth day after injection. For each chlorophyll assay, the leaves of the plant were removed from the stem. For each 0.1 g of leaves, 6.0 mL of 100% methanol were used. The leaves were thoroughly ground in half of the methanol with a pestle in a mortar. The leaves were ground again after the rest of the methanol was added. Extraction of the chlorophyll was allowed to proceed for 45 min at room temperature. Then the suspension was gravity filtered through filter paper to remove the leaf parts. The absorbance of the filtrate was measured with a Spectronic 20 spectrophotometer at 652 nm and 665.2 nm. The absorbance values were converted to relative chlorophyll units using the following equation derived by Porra and colleagues (1989):

Total chlorophyll (a and b) = Dilution factor × $[22.12 A_{652\,nm} + 2.71 A_{665.2\,nm}$ (mg/L)] × Volume of solvent (L) / Weight of leaves (mg)

Make proper subscripts.

Results
P. infestans-treated plants and the control plants had similar growth patterns (Figure 2). Both the experimental and control pea and corn plants grew at a

Include text in the Results section. Describe the important results shown in each figure and table.

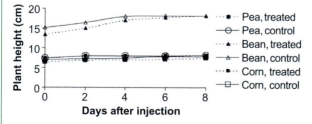

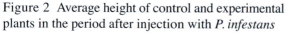

Figure 2 Average height of control and experimental plants in the period after injection with *P. infestans*

Refer to each figure and table in parentheses.

Make text in legend and in axes titles large enough to read easily.

Make sure intervals on axes have correct spacing.

Make points and lines black and background white for best contrast.

In the axes titles, write the variable followed by the units in parentheses (where applicable).

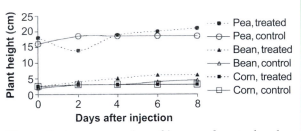

Figure 3 Average number of leaves of control and experimental plants in the period after injection with *P. infestans*

Position the figure caption below the figure.

constant, but very slow rate over the eight day test period. The control bean plants were taller on average than the experimental bean plants throughout most of the experiment. Both groups showed the same growth pattern, however, with rapid growth occurring from day 18 to 24 (0 to 4 days after injection), followed by slower growth to the end of the experiment.

Check that figures are numbered in sequence.

As plant height increased, the average number of leaves on all of the plants also increased over the measurement period (Figure 3). There is an uncharacteristic decrease in the number of leaves of pea plants treated with *P. infestans* from day 18 to 20 (0 to 2 days after injection), but this is probably due to counting error.

Describe the figures in order.

There was a general decline in average leaf angle of all the plants over the first four days after injection with *P. infestans* (Figure 4). The plants did not follow this pattern over the second half of the experiment, however. The leaf angle of the experimental bean group increased by 28°, while that of the control bean group only increased by about 3°. The leaf angle of the control pea plants increased significantly (33°), while that of the experimental pea plants decreased 4°. The leaf angle of the corn control group decreased 0.5°, while that of the corn experimental group showed a much sharper decline of 24°.

Insert symbols such as ° using word processing software.

There was also no difference between the experimental and control groups with regard to chloro-

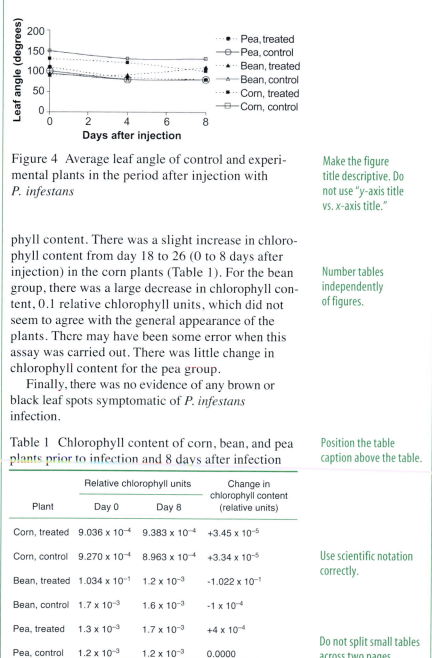

Figure 4 Average leaf angle of control and experimental plants in the period after injection with *P. infestans*

Make the figure title descriptive. Do not use "y-axis title vs. x-axis title."

phyll content. There was a slight increase in chlorophyll content from day 18 to 26 (0 to 8 days after injection) in the corn plants (Table 1). For the bean group, there was a large decrease in chlorophyll content, 0.1 relative chlorophyll units, which did not seem to agree with the general appearance of the plants. There may have been some error when this assay was carried out. There was little change in chlorophyll content for the pea group.

Number tables independently of figures.

Finally, there was no evidence of any brown or black leaf spots symptomatic of *P. infestans* infection.

Table 1 Chlorophyll content of corn, bean, and pea plants prior to infection and 8 days after infection

Position the table caption above the table.

Plant	Relative chlorophyll units		Change in chlorophyll content (relative units)
	Day 0	Day 8	
Corn, treated	9.036×10^{-4}	9.383×10^{-4}	$+3.45 \times 10^{-5}$
Corn, control	9.270×10^{-4}	8.963×10^{-4}	$+3.34 \times 10^{-5}$
Bean, treated	1.034×10^{-1}	1.2×10^{-3}	-1.022×10^{-1}
Bean, control	1.7×10^{-3}	1.6×10^{-3}	-1×10^{-4}
Pea, treated	1.3×10^{-3}	1.7×10^{-3}	$+4 \times 10^{-4}$
Pea, control	1.2×10^{-3}	1.2×10^{-3}	0.0000

Use scientific notation correctly.

Do not split small tables across two pages.

Discussion

P. infestans did not affect the plant height, leaf angle, number of leaves, and chlorophyll content of *Zea mays, Pisum sativum,* or *Phaseolus.* Symptoms of infection are the presence of brown or black spots (areas of dead tissue) on leaves and stems, and, as the infection spreads, the entire plant becomes covered with a cottony film (Stanley, 1994). None of the experimental plants exhibited these symptoms.

There may be several reasons why *P. infestans* did not affect the plants in this study. One reason is that the L-broth culture of *P. infestans* may not have contained zoospores of the fungus. Zoospores are motile spores that can penetrate the host plant through the leaves and soft shoots, or through the roots (Stanley, 1994). Zoospores are usually produced in wet, warm weather conditions (Ingold and Hudson, 1993). If the L-broth culture did not contain any zoospores, or if the soil around the plants was not sufficiently saturated to stimulate production of zoospores, then these conditions may have prevented *P. infestans* from attacking the roots and shoots of the plants.

In order to determine if the problem was lack of zoospores, first the L-broth culture could be examined microscopically for presence of zoospores. Second, the *P. infestans* plants could be watered with different quantities of water to determine if the fungus requires wetter soil for zoospore production and motility.

Another reason why *P. infestans* may not have affected the plants is that this species of fungus may be specific to potato (*Solanum tuberosum*) and tomato (*Lycopersicon esculentum*) plants (Stanley, 1994), which both belong to the nightshade family (Solanaceae). In contrast, corn belongs to the grass family (Gramineae), and peas and beans are legumes (Leguminosae). It may be that these plant families are not susceptible to *P. infestans*, which has a very limited host range (Stanley, 1994). Non-susceptible plants have been shown to have defense mechanisms that prevent *P. infestans* from infecting them (Gallegly, 1995).

Further research is required to determine if *P. infestans* really cannot infect corn, pea, and bean plants. Goth and Keane (1997) developed a test to measure resistance of potato and tomato varieties to original and new strains of *P. infestans*. Their experiments involved exposing the experimental plants' leaves directly to the fungus, and this method could perhaps be tested on corn, pea, and bean leaves as well.

References

Brave New Potato. 1994. Discover 15(10): 18–20.

Gallegly ME. 1995. New criteria for classifying Phytophthora and critique of existing approaches. In: Erwin DC, Bartnicki-Garcia S, Tsao PH, editors. Phytophthora: Its Biology, Taxonomy, Ecology, and Pathology St. Paul: The American Phytopathological Society. pp. 167–172.

Goth RW, Keane J. 1997. A detached-leaf method to evaluate late blight resistance in potato and tomato. American Potato Journal 74(5): 347–352.

Ingold CT, Hudson HJ. 1993. The Biology of Fungi, 6th ed. London: Chapman and Hall.

McElreath, Linda R. 1994. One potato, two potato. Agricultural Research 42(5): 2–3.

Porra RJ., Thompson WA, Kriedemann PE. 1989. Determination of accurate extinction coefficients and simultaneous equations for assaying chlorophylls a and b extracted with four different solvents: verification of the concentration of chlorophyll standards by atomic absorption spectroscopy. Biochimica et Biophysica Acta 975: 384–394.

Stanley D. 1994. What was around comes around. Agricultural Research 42(5): 4–8.

Stanley D. 1997. Potatoes. Agricultural Research 45(5): 10–14.

In Name-Year end reference format, list authors alphabetically by first author's last name.

Substitute the title when no author is given.

Use mostly primary journal articles or articles in a book.

List all authors (up to 10; then list first 10 followed by *et al.* or *and others*).

See the tabbed pages in Chapter 4 for examples of how to reference printed and electronic sources.

Give inclusive page numbers, not just the page(s) you extracted information from.

Make sure all in-text citations have a corresponding end reference.

Make sure all end references have a corresponding in-text citation.

Laboratory Report Mistakes

The following table lists some common mistakes made by students writing biology lab reports. When you proofread the first draft of your report, look over this list and be on the lookout for these kinds of mistakes. This table can be printed from <http://sites.sinauer.com/Knisely4E> .

Your instructor may also use this key to save time grading lab reports without sacrificing the quality of the feedback. For example, the number "8" written in the left margin means that you failed to include essential information in the abstract. If your instructor uses this system, be sure to refer to the key for an explanation of your instructor's comments.

Although some of these mistakes may not affect the content of your paper, they do affect your credibility as a scientist; careless writing may be equated with careless science.

MISTAKE	EXPLANATION
lc	Use lowercase letter
CAP or uc	Use capital (uppercase) letter
^	Insert text
⟩	Delete text
⌡	Leave space between the two characters
⌢⌣	Close up
¶	Start a new paragraph
Page break	End the current page; move subsequent text to next page. (In Microsoft Word and WordPerfect, press Ctrl + Enter where you want to end the page.)
agr	Subject and verb do not agree.
wc	Word choice. Word used is not appropriate for the situation.
. . . . . .	A dotted underline means "stet," or "let original text stand." The correction was made in error.
1	Word usage incorrect. Look it up in the dictionary. Examples:
a.	absorbance (how much light is absorbed) vs. absorbency (how much moisture a diaper or paper towel can hold)
b.	to bind (what atoms and molecules do) vs. to bond (what glue does)
c.	complementary (DNA strands) vs. complimentary (given free as a courtesy)

d.	data = plural of "datum"; e.g. "These data show…" not "This data shows…"
e.	effect (noun) vs. effect (verb meaning "to cause") vs. affect (verb meaning "to influence")
f.	conformation (3-D shape of a protein) vs. confirmation (verification)
i.	it's (it is) vs. its (belonging to "it")
m.	amount (cannot count individual pieces) vs. number (can count individual pieces)
r.	ratio (proportion or quotient) vs. ration (a fixed portion of food)
s.	strain of bacteria vs. strand of DNA
t.	then (next in time) vs. than (an expression for comparing two things)
v.	varying (changing) vs. various (different)
2	Italicize Latin names.
3	Write in passive voice. Shift the emphasis from yourself to the subject of the action.
4	Use punctuation correctly.
5	Do not write awkward or incomplete sentences.
6	Do not write symbols by hand. Use **Insert	Symbols	Symbol** on the Ribbon.
7	Superscript/Subscript.
	Superscript: For 10^{-5}, for example, type "10-5"; then select "-5" and click **Home	Font	Superscript.**
	Subscript: For K_m, for example, type "Km"; then select "m" and click **Home	Font	Subscript.**
8	Essential abstract content is missing. Include purpose, brief description of methods, results, and conclusions or possible explanation for the results.
9	Too much detail for abstract.
10	Cite sources when providing background.
11	State the purpose of the current experiment.
12	Wrong citation format. Use Name-Year, Citation-Sequence, or Citation-Name format. See Chapter 4 for specific examples.
13	Do not use direct quotations in scientific writing. Paraphrase and cite the source.

Laboratory Report Mistakes

14	Write Materials and Methods section in past tense and in paragraph form. Do not make a numbered list.
15	Do not list materials separately in Materials and Methods section, unless the source is noncommercial or critical for the outcome of the experiment.
16	Time frame does not affect the outcome.
17	Do not refer to the container; instead emphasize the contents and give appropriate volumes, masses, and concentrations.
18	Do not describe routine laboratory procedures in detail.
19	In the Materials and Methods section, do not give a "preview" of how data will be organized in the Results section.
20	Essential details are missing in the Materials and Methods section. Provide enough detail to enable the reader to repeat the experiment.
21	Do not include raw data in the Results section; instead, summarize and organize the data.
22	Use *past* tense when describing your own results. Present tense is reserved for information already accepted by the scientific community.
23	Describe *each* table and *each* figure individually and sequentially.
24	Refer to each figure/table number in parentheses as you describe the results.
25	Do not use unnecessary introductions like "Fig. 1 shows the effect of *x* on *y*." Instead, describe *how x* affected *y*.
26	Do not include a table when the figure(s) shows the same data.
27	Give each figure and table a number and a title.
28	Use figure and table titles that are self-explanatory without referring to the text.
29	Do not write uninformative figure titles such as "*y*-axis title vs. *x*-axis title." See Chapter 4 for examples.
30	Position the figure caption *below* the figure; position the table caption *above* the table.
31	Do not include a title above the figure. When you make the figure in Excel, leave the "Chart title" space blank.
32	Use proper spacing on the axis. Choose "XY Scatter" plot in Excel (see Appendix 2).
33	Give each data set its own distinguishable symbols and lines. Choose black for both the line and the symbol color.

34	Use scientific notation when numbers are very large or very small.
35	Include units.
36	Use correct number format. Do not start sentences with a number (write out the number). Make sure decimal numbers less than one start with zero, e.g., 0.1 mL, *not* .1 mL.
37	Give possible explanations for the results.
38	Compare your results with those in the literature. Science is not done in a vacuum!
39	Use proper reference format. See Chapter 4 for specific examples.
40	List all in-text citations in the References section.
41	Cite all end references in the text.

Laboratory Report Mistakes

Chapter 7

POSTER PRESENTATIONS

Posters are a means of communicating research results quickly. They provide a great opportunity to get feedback about preliminary data and ideas. **Poster sessions** are often held at large national meetings, and they allow you to meet other scientists in an informal setting.

Why Posters?

Scientists who attend poster sessions constitute a much larger audience than the one attracted to a journal article on a particular topic. Thus, your goal is to produce a poster that not only attracts experts in your subdiscipline, but also the much larger group of scientists with tangential research interests. The latter group provides a unique opportunity for you to learn about applications of your work to other research areas (and vice versa), spurs scientific creativity, and prompts you to apply an interdisciplinary approach to problem-solving.

Posters are *not* papers; they rely more on visuals than on text to present the message. It is not necessary to supply as many supporting details as you would for a paper, because you (the author) will be present to discuss details one-on-one with interested individuals. Too much material may even discourage individuals from reading your poster.

An appropriate poster presentation should fulfill two objectives. First, it must be esthetically pleasing to attract viewers in the first place. Second, it must communicate the objectives, methods, results, and conclusions clearly and concisely.

Poster Format

Posters can be constructed on poster boards or created in PowerPoint and printed on large-format paper. If you are presenting at a professional society meeting, check with the conference organizer regarding size and other requirements. For poster sessions in your class, ask your instructor about appropriate materials and sizes.

Layout

Make a rough sketch of the layout before you begin. Position the title and authors prominently at the top of the poster. Then arrange the body of the poster in 2, 3, or 4 columns, depending on whether the orientation of the poster is portrait or landscape.

Appearance

The success of a poster presentation depends on its ability to attract people from across the room. Interesting graphics and a pleasant color scheme are good attention-getters, but avoid "cute" gimmicks. Present your poster in a serious and professional manner so people will take your conclusions seriously.

Organize the sections so that information flows from top left to bottom right. Align text on the left rather than centering it. The smooth left edge provides the reader with a strong visual guide through the material.

Avoid crowding. Large blocks of text turn off viewers; instead, use bullets to present your objectives and conclusions clearly and concisely. Use blank space to separate sections and to organize your poster for optimal flow from one section to the next.

Use appropriate graphics that communicate your data clearly. Use three-dimensional graphs *only* for three-dimensional data. Proofread figure legends and running text carefully and correct spelling and grammatical errors.

Colored borders around graphics and text enhance contrast, but keep framing to a minimum. **Framing** is the technique whereby the printed material is mounted on a piece of colored paper, which is mounted on a piece of different-colored paper to produce colorful borders. Use borders judiciously so that they do not distract from the poster content.

When using poster boards, use adhesive spray or a glue stick to affix text and figures. These tend to have fewer globs and bulges than liquid glues.

Font (type size and appearance)

Remember that most readers of your poster will be 3 to 6 feet away, so the print must be large and legible. Sans serif fonts like Arial are good for titles, but serif fonts like **Times** and **Palatino** are much easier to read in extended blocks of text. The **serifs** (small strokes that embellish the character at the top and bottom) create a strong horizontal emphasis, which helps the eye scan lines of text more easily.

Make the title of the poster title case or ALL CAPS in 72 point **bold**. In **title case**, the most important words are capitalized. Title case has the advantages of being easier to read and taking up less space than all caps. Do not use all caps if there are case-sensitive words in the title, such as pH, cDNA, or mRNA. Limit title lines to 65 characters or less.

Authors' names and affiliations should be 48 or 36 point **bold**, serif font, title case:

Times 48 Pt

The section headings can be 28 point **bold**, serif font, title case:

Times 28 Pt

The text itself should be no smaller than 24 point, serif font, sentence case, and *not* in bold:

Times 24 pt

Nuts and bolts

Ask the conference organizer (or your instructor) about how posters will be displayed at the poster session. Some possibilities include a pinch clamp on a pole, an easel, a table for self-standing posters, and cork bulletin boards to which posters are affixed with pushpins.

Making a Poster in Microsoft PowerPoint 2010

Many scientists use PowerPoint to prepare posters for professional society meetings because this software is readily available; the font, color scheme, and layout can be customized by the authors; the content can be revised easily; the electronic file format facilitates collaboration among colleagues in different locations; and the final product looks professionally done. Because you are at the mercy of your computer when making posters in PowerPoint, however, save your file early and often!

The entire poster is made on one PowerPoint slide, whose size is set in the **Page Setup** dialog box. Text is written in textboxes, which are inserted along with images and graphs on the slide as desired. The following sections describe some of the basic tasks you will carry out when making your poster in PowerPoint 2010.

Page setup

1. Click **Home | Slides | Layout** and choose **Blank**.
2. Click **Design | Page Setup | Page Setup** and enter the poster dimensions (Figure 7.1). If you are using plotter paper, the maximum size is about 34 × 56". Portrait or landscape orientation will be selected automatically based on the numbers you enter.
3. If desired, select a theme or a color for the background on the **Design** tab.

Adding text, images, and graphs

- To add text to your poster, click **Insert | Text | Textbox** for each section or block of text. Aim for a consistent look by using the same family of fonts. Adjust the font size for the title,

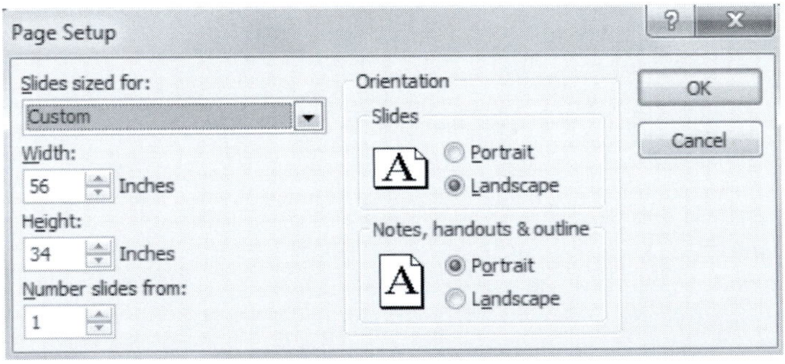

Figure 7.1 Poster dimensions are specified in the **Page Setup** dialog box on the **Design** tab.

authors' names and affiliations, section headings, and text by making the appropriate selections on **Home | Insert | Font**. To change the properties of the textbox itself, right-click it and select **Format Shape**.

- To insert images, click **Insert | Picture**.
- To insert a graph, copy and paste it from Excel or another graphing program. Right-click the Chart Area to enlarge the font to 24 pt. Right-click the axes and data sets to make the lines thicker.
- To make your own graphics, see "Line Drawings" in Appendix 3.
- Use the **Zoom slider** in the lower right corner to enlarge the section of the poster you're working on.

Aligning objects

1. To center the title and authors on the slide, hold down the **Shift** key, click each textbox, and click **Arrange | Align | Align Center** on the **Drawing Tools Format** tab (Figure 7.2).
2. Now display the ruler by clicking **View | Show | Ruler**. Drag or nudge the grouped textboxes so that the center handle lines up with the zero on the horizontal ruler.
3. To align the left edges of textboxes and images, hold down the **Shift** key, click each object, and then click **Align Left** on the **Drawing Tools Format** tab.

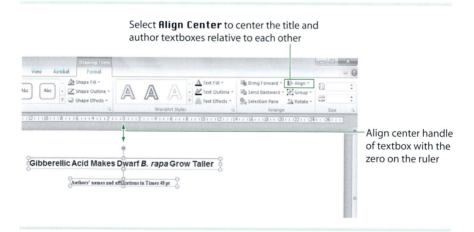

Select **Align Center** to center the title and author textboxes relative to each other

Align center handle of textbox with the zero on the ruler

Figure 7.2 Center the title and authors names and affiliations by first clicking **Align Center** and then aligning the center handle of the textbox with the zero on the horizontal ruler.

Proofread your work

Print a rough draft of your poster on an 8.5 × 11″ sheet of paper. To do so, click **File | Print**. Under **Settings**, click the down arrow next to **Full Page Slides** and select **Scale to Fit Paper** under **Options**. Check for typos and grammatical errors and evaluate the layout and overall appearance of your poster. If the print is not large enough to read on the draft, it will probably be hard to read from a distance on the full-sized poster as well. Adjust the font size under **Home | Font**.

Final printing

Check with your local print shop concerning the electronic format of the poster file. PowerPoint (.ppt or .pptx) may be acceptable, but Portable Document Format files (PDFs) are becoming the printing industry standard. PDFs tend to be smaller than the source file and all of the fonts, images, and formatting are retained when the document is printed.

Poster Content

Posters presented at professional society meetings should be organized so that readers can stand 10 feet away from the poster and get the take-home message in 30 seconds or less. Because of the large number of sessions (lectures) and an even larger number of posters, conference participants often experience "sensory overload." Thus, if you want your poster to stand out, make the section headings descriptive, the content brief and to the point, and the conclusions assertive and clear.

Posters for a student audience in the context of an in-house presentation should follow the same principle of brevity, but may retain the sections traditionally found in scientific papers. These include:

- Title banner
- Abstract (optional—if present, it is a summary of the work presented in the poster)
- Introduction
- Materials and Methods
- Results
- Discussion or conclusions
- References (less comprehensive than in a paper)
- Acknowledgments

Title banner

Use a short, yet descriptive, title. This is the first and most important section for attracting viewers, so try to incorporate your most important conclusion in the title. For example, **Gibberellic Acid Makes Dwarf B. *rapa* Grow Taller** is more effective than **Effect of Gibberellic Acid on Dwarf B. *rapa*.**

The title banner should be at the top of the poster and in 72 point bold font, title case, or all caps, 65 characters or less on a line. Underneath the title, include the authors' names and the institutional affiliation(s). Use 48 point bold, title case for the authors' names.

Introduction

Instead of using the conventional "Introduction" heading found in papers, consider using a short statement of the topic or introduce the topic as a question. In this section, briefly explain the existing state of knowledge of the topic, why you undertook the study, and what specifically you intended to demonstrate. A bulleted list of objectives may be a good way to present some of this information.

Materials and methods

Present the methods you used to investigate the problem in enough detail so that someone competent in basic laboratory techniques could repeat your experiments. You might write the basic approach as a series of bulleted statements, and then provide more details in the subsequent text. Be both *brief* and *thorough*.

Results

The Results section of a poster consists mostly of visuals (images and graphs) and a minimum of text. Poster viewers do not have time to read the results leisurely, as they do with a journal article. An effective presentation of the results should announce each result with a heading, provide a visual that displays the result, and use text sparingly as a supplement to the visuals.

Graphs should summarize the raw data in a manner that allows viewers to appreciate both the general patterns of the data and the degree of variability that they possess. Written text should concentrate on general patterns, trends, and differences in the results, and not on the numbers themselves. For example, the reader can visualize "the concentration of chlorophyll increased initially, and then leveled off after 10 days" much more easily than "the chlorophyll concentration was 9×10^{-5} units on the fourth day, increased to 6×10^{-4} units on the tenth day, and then stayed about the same at 4×10^{-4} units on day 21."

Avoid using tables with large amounts of data; if you think the data are important, prepare the table to give as a handout. Your job is to sort through the data and come up with the take-home message. Do not use flip-out charts, in which one table or figure is displayed beneath another.

Figures may contain some statistical information including means, standard error, and minimum and maximum values, where appropriate. Make the data points prominent and use a simple vertical line without crossbars for the error bars. Use the same sized font for the axes labels and the key as you use for the text (24 pt). Do not present the same data as both a table and a figure.

Images are a great addition to a poster, but only if they are sharp. Picture quality is determined by the size (number of pixels) of the digital image and the printer resolution. Various sources recommend a printer resolution of at least 240 pixels per inch (ppi) to print quality photos. That means for a $5 \times 7"$ print, the image would have to be at least 1200×1680 pixels to attain the desired resolution on paper. To check the size of an image, right-click it and select **Properties**. Be careful when resizing pictures. Enlarging an image that does not have enough pixels will result in a blurry picture.

Visuals in posters do not need a caption with the number and title. Instead, the graphs and pictures are integrated so that they immediately follow the text in which they are first described.

Edit the text ruthlessly to remove nonessential information about the visuals. A sentence like "The effect of gibberellic acid on *B. rapa* is shown in the following figure" is nothing but deadwood. On the other hand, "*B. rapa* plants treated with gibberellic acid grew taller than those receiving only water" informs the reader of the result.

Remember to leave some blank space on a poster. Space can be used to separate sections and gives the eyes a rest.

Discussion or conclusions

Interpret the data in relation to the original objective or hypothesis and relate these interpretations to the present state of knowledge presented in the introduction. Discuss any surprising results. Discuss the future needs or direction of the research. Where appropriate, identify sources of error and basic inadequacies of the technique. Do not cover up mistakes; instead, suggest ways to improve the experiment if you were to do it again. You may also speculate on the broader meaning of your conclusions in this section. Use bullets to help present this information concisely.

Literature citations

In posters as in scientific papers, it is common to cite the work of others, particularly in the Introduction and Discussion sections. Because posters are informal presentations and the author is present to provide supporting details, the References section of a poster can be less comprehensive than that in a paper. Scientists who visit your poster are likely to work in the same field and probably are already familiar with most of the literature, so a long list of references would only waste valuable space on your poster. Even visitors who have a general knowledge of your topic, but who work in a different subdiscipline, are not interested in the details. They are mainly interested in how your approach or findings might help them improve their methodology or provide insight into their work. Discussions with scientists in this second group are valuable to you because they provide a different perspective and may help you see applications to other subdisciplines.

Student presentations in an in-house setting are different from poster sessions at large national meetings because students usually do not have the background or the familiarity with the literature that career researchers have. Researching your topic is part of the scientific method, however, and presenting your work in the context of the published literature is part of good science. Thus, your instructor may ask you to include literature citations on your poster or to provide a list of references on a handout. See the section "Documenting Sources" in Chapter 4.

Acknowledgments

In this section, the author acknowledges organization(s) that provided funding and thanks technicians, colleagues, and others who have made significant contributions to the work.

Presenting Your Poster

Authors should prepare a 5-minute talk explaining their poster; anticipate questions from students and instructors, and prepare appropriate answers. After this presentation, at least one author must be present at his or her poster at all times to answer questions from the session participants. The more you interact with your audience, the more feedback you are likely to receive on your work. On the other hand, do not throw yourself on participants who demonstrate little interest.

Evaluation Form for Poster Presentations

Good posters are the product of creativity, hard work, and feedback at various stages of the poster preparation process. When you present a poster to your class or at a professional society meeting, participants may be asked to evaluate your poster using a form such as the "Evaluation Form for Poster Presentations" available at <http://sites.sinauer.com/Knisely4e>.

When you are in the position of evaluator, make the kinds of comments you would find helpful if you were the presenter. As you know, feedback is most likely to be appreciated when it is constructive, specific, and done in an atmosphere of mutual respect.

Sample Posters

Examples of posters along with reviewer comments are posted on <http://sites.sinauer.com/Knisely4e>. What do you like (or not like) about each poster? Use the evaluation form to determine how well the authors have met the requirements of good poster design.

Chapter 8

ORAL PRESENTATIONS

Scientific findings are communicated through journal articles, poster presentations at meetings, and oral presentations. Oral presentations are different from journal articles and posters because the speaker's delivery plays a critical role in the success of the communication.

As a student, you have been on the receiving end of oral presentations for a number of years and probably have a pretty good idea of what makes a talk enjoyable. For example, presentations in which the speaker appears to be unprepared, uncomfortable speaking in public, and unfamiliar with the level of his or her audience are quickly forgotten regardless of the content. On the other hand, memorable presentations tend to have the following characteristics:

- Speakers who establish a good rapport with the audience.
- Information that is well organized and leaves listeners satisfied that they understand more than they did before.
- Visuals that are simple and legible, which help listeners focus on the important points without drawing attention away from the speaker.

Organization

Oral presentations are not unlike scientific papers in their structure, but they are much more selective in their content. As in a scientific paper, the **introduction** captures the audience's attention, provides background information on the subject, and clarifies the objectives of the work (Table 8.1). The **body** is a condensed version of the Materials and Methods and Results sections, and contains only the details necessary to emphasize and support the speaker's conclusions. If the focus of the talk is on the results, then the speaker spends less time on the methods and more time on visuals that high-

TABLE 8.1	Comparison of the structure of an oral presentation and a journal article	
Oral Presentation	**Journal Article**	
	Abstract	
Introduction	Introduction	
Body	Materials and Methods, Results	
Closing	Discussion	
	References	
	Acknowledgments	

light the findings. The visuals may be simpler than those prepared for a journal article, because the audience may only have a minute or two to digest the material in the visual (in contrast to a journal article, where the reader can re-read the paper as often as desired). Finally, the **closing** is comparable to the Discussion section, because here the speaker summarizes the objectives and results, states conclusions, and emphasizes the take-home message for the audience. As part of the closing, you may include an acknowledgments slide in which you recognize sources of funding as well as your research adviser and others who helped you with your work.

There is no distinct abstract or References section in an oral presentation. If the presentation is part of a meeting, an abstract may be provided in the program. If listeners are interested in the references, they should contact the speaker after the talk.

Plan Ahead

Before you start writing the content of your presentation, make sure you know the following:

- Your audience. What do they know? What do they want to know?
- Why you are giving the talk, not just why you did the experiment. Are you informing the audience, or are you trying to persuade them to adopt your point of view?
- Your speaking environment. How much time do you have? How large is the audience? What audio-visual equipment is available (chalkboard, overhead projector, computer with projection equipment)?

Write the Text

An oral presentation will actually take you longer to prepare than a paper. D'Arcy (1998) breaks down the steps as follows:

- Procrastination   25%
- Research   30%
- Writing and creating visuals   40%
- Rehearsal   5%

To overcome procrastination, follow the same strategy as for writing a laboratory report: write the body first (Materials and Methods and Results sections), then the introduction, and finally the closing.

- Write the main points of each section in outline form.
- Use the primary literature (journal articles) to provide background in the introduction and supporting references in the closing.
- Transfer your outline to notecards. Use only one side of the notecards, and limit yourself to one main topic per card. Note where you plan to use visual aids.
- If you plan to use a presentation graphics program such as Microsoft PowerPoint in addition to notecards, make sure each slide is just an *outline* of what you're going to say. Use the slides to *support* what you say, not *repeat* what you say.
- Keep it short and simple. Tell your audience what you're *going* to tell them, *tell* them, and then tell them what you *told* them.

Prepare the Visuals

Effective visual aids help the audience to visualize the data and to see relationships among variables. They clarify information that would otherwise involve lengthy descriptions.

Effective visuals are simple, legible, and organized logically. They capture the listener's interest, increase comprehension, and focus attention on the important findings. When deciding on which visuals to use, keep in mind that student audiences prefer figures to tables. For guidelines on what kind of figure to use, see "Preparing visuals" and "Organizing your data" in Chapter 4. For instructions on how to make XY graphs in Microsoft Excel, see Appendix 2. Photos, diagrams, and drawings are effective ways to show physical rather than numerical results. Mechanisms are much easier to visualize when the verbal description is accompanied by a drawing.

Take the time to make good visuals for your oral presentation. We receive more than 83% of our information through sight, and only 11% through hearing. Without visuals, your listeners will forget most of your presentation within 8 hours. These statistics help explain why Microsoft PowerPoint has become so popular in recent years. PowerPoint allows you to create a slide show containing text, graphics, and even animations (see Appendix 3). PowerPoint has the same advantages as word processing programs in that it is easy to make revisions and even send an electronic copy of the presentation to a colleague to review.

Because PowerPoint is so easy to use, you may be tempted to write every word of your presentation on the slides. Don't do it. Follow these tips for preparing visuals:

- **Do not put too much information on any one slide or transparency.** A good rule of thumb is to cover only one concept per slide or transparency; use overlays for more complicated concepts that require more than one slide.

- **Keep the wording simple.** Instead of using full sentences, bullet keywords or key points (Figure 8.1). Limit the number of key points to 3–6.

- **Make sure the lettering can be read from anywhere in the room.** Use no smaller than 18–24 pt font and black type on a white background for best contrast. The light background makes the room much brighter, which has the dual advantage of keeping the audience awake and allowing you to see your listener's faces. Use color sparingly in text slides; a bright color

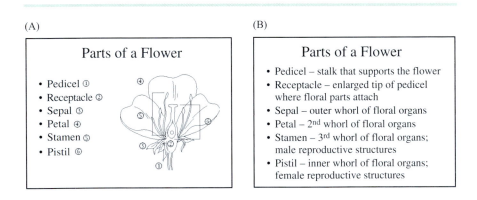

Figure 8.1 Effective (A) and less effective (B) ways to use text on a slide. The wording in Slide A is simple and the numbered reference to the figure makes it easy for viewers to visualize the location of each floral part. The large amount of text on Slide B will have the audience reading instead of listening to the speaker, and the absence of a diagram slows comprehension.

or a larger font, when used with discretion, are good ways to emphasize a point.

- **Use the same font and format for each slide for a consistent look.** Avoid "cute" graphics that might detract from the scientific message.

- **Keep graphs simple.** Make the numbers and the lettering on the axes labels large and legible. Choose symbols that are easy to distinguish. Instead of following the conventions used in journal articles, modify the figure format as follows (Figure 8.2):

(A)

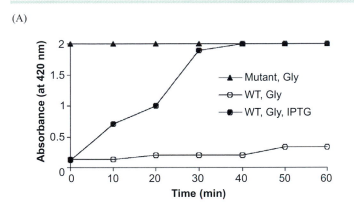

Figure 1   Beta-galactosidase production in *E. coli* grown under three different conditions. The mutant strain, which produces a nonfunctional repressor protein, was grown in glycerol. The wild-type strain was grown in glycerol, with or without the addition of IPTG (a lactose analog) at *t* = 0 min.

(B)

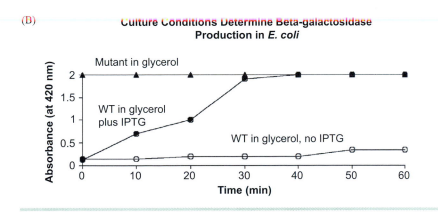

Figure 8.2   Example of figure formatting for (A) a journal article and (B) an oral presentation.

- Position the title above the figure and don't number it. The title should reflect your most important finding.
- Label each line instead of identifying the data symbols with a key. The labels may be made using textboxes in Excel, Word, or PowerPoint.

Both of these modifications make it easier for the listener to digest the information within the short time the graph is displayed during the presentation.

- **Use symbols consistently in all of your graphs.** For example, if you use a triangle to represent "Mutant in glycerol" in the first graph, use the same symbol for the same condition in the rest of your graphs.
- **Appeal to your listeners' multiple senses.** Use images from the Internet, recordings of heart sounds or bird songs, and animations or movie clips if they help make your point. Make sure these "extras" do not detract from your take-home message and that they fit within the time limit of your presentation.
- **Allow at least 20 seconds for each slide, more time for complicated slides.**

Rehearsal

After you have written your presentation and prepared your visuals, you must practice your delivery. Give yourself plenty of time so that you can run through your presentation several times and, if possible, do a practice presentation in the same room where you will hold the actual presentation. Here are some other tips:

- Go to a place where you can be alone and undisturbed. Read your presentation aloud, paying attention to the organization. Does one topic flow into the next, or are there awkward transitions? Revise both the PowerPoint and your speaker notes as needed. Apply the revision strategies described in Chapter 5 to your oral presentation.
- Time yourself. Make sure your presentation does not exceed the time limit.
- After you are satisfied with the organization, flow, and length, ask a friend to listen to your presentation. Ask him or her to evaluate your poise, posture, voice (clarity, volume, and rate), gestures and mannerisms, and interaction with the audience (eye contact, ability to recognize if the audience is following your talk).

It is natural to be nervous when you begin speaking to an audience, even an audience of your classmates. Adequate knowledge of the subject, good preparation, and sufficient rehearsal can all help to reduce your nervousness and enhance your self-assurance.

Delivery

The importance of the delivery cannot be overstated: listeners pay more attention to body language (50%) and voice (30%) than to the content (20%) (Fegert and others 2002). Remember that you must establish a good rapport with the audience in order for your oral presentation to be successful. The following guidelines will help.

Presentation style

- Number your notecards so that if you drop them you can reassemble them quickly.
- Dress appropriately for the occasion.
- Use good posture. Good posture is equated with self-assurance, while slouching implies a lack of confidence.
- Be positive and enthusiastic about your subject.
- Try to maintain eye contact with some members of the audience. Eye contact is perceived as more personal, as though the speaker is talking with individual listeners.
- Avoid distracting gestures and mannerisms such as pacing, fidgeting with the pointer, jingling the change in your pocket, and adjusting your hair and clothes.
- Speak clearly, at a rate that is neither too fast nor too slow, and make sure your voice carries to the back of the room. Avoid "um's," "ah's," and other nervous sounds.

Integrating visuals

Visuals are an integral part of an oral presentation. If you use **transparencies**, be sure to:

- Number transparencies in chronological order. If they fall down, you can reassemble them easily.
- Avoid pointing your finger on the transparency itself (your finger will be a blur on the projection screen). Instead lay a pen on the transparency as a pointer, or point directly to the screen.

If you use **PowerPoint** to prepare your presentation, do not stand behind the computer monitor and read your presentation off the screen. This practice distances you from your audience.

No matter which method you choose, the following points apply:

- Do not display a visual until you are ready to discuss it.
- Stand next to the projection screen and point with your outstretched arm or pointer (stick) or use a laser pointer.
- Tell the audience exactly what to look for. Interpret statistics and numbers so that they are meaningful to your listeners. If you are describing Figure 8.2, for example, "Beta-galactosidase production increased dramatically in only 10 minutes after adding the lactose analog" is likely to make a bigger impression on the audience than "The absorbance increased from 0.12 to 0.71 in only 10 minutes after adding the lactose analog."
- Point to the specific part of the visual that illustrates what you are describing. Avoid waving the pointer in the general direction of the visual.
- Do not block the projected image with your body.
- Do not turn your back on the audience.

Interacting with the audience

A unique benefit of communicating information verbally is that you can adjust your presentation based on the feedback you receive from your listeners. Pay attention to the reaction of individuals in the audience. Does their posture or facial expression suggest interest, boredom, or confusion? If you perceive that the audience is not following you, ask a question or two, and adjust your delivery according to their response. Your willingness to customize your presentation to your audience will enhance your reputation and lead to greater listener satisfaction.

To keep listeners focused:

- Establish common ground immediately. Ask a question or describe a general phenomenon with which your audience is familiar.
- Make sure your introduction is well organized and proceeds logically from the general to the specific without any sidetracks.
- Alert your audience when you plan to change topics. Practice these transitions when you rehearse.
- Summarize the main points from time to time. Tell your listeners what you want them to remember.

- Use examples to illustrate your hypotheses and conclusions.
- End your presentation with a definite statement such as "In conclusion…" Don't fade out weakly with phrases such as "Well, that's about it."

Group presentations

If you are presenting with another person, your individual contributions should complement one another, not compete with each other. This requires good coordination and practice beforehand.

Fielding listener questions

Allow time for questions and discussion. Take questions as a compliment—your listeners were paying attention! Listen carefully to each question, repeat it so that everyone in the room can hear, and then give a brief, thoughtful response. If you don't understand the question, ask the listener to rephrase it. If you don't know the answer, say so.

Signal the end of the question session with "We have time for one more question" or something similar.

Feedback

It takes not only practice, but also coaching, to become an effective speaker. When you make an oral presentation to your class, your instructor may ask your classmates to evaluate your delivery, organization, and visual aids, as well as the thoroughness of your research using a form such as the "Evaluation Form for Oral Presentations" available at <http://sites.sinauer.com/Knisely4e>.

When you are in the position of the listener, make the kinds of comments you would find helpful if you were the speaker. As you know, feedback is most likely to be appreciated when it is constructive, specific, and done in an atmosphere of mutual respect.

WORD PROCESSING IN MICROSOFT WORD 2010

Introduction

This appendix deals exclusively with Microsoft Word 2010 using the Windows 7 operating system. While there are similarities between the Word 2010 and Word for Mac 2011 Ribbon interfaces, some of the commands are accessed differently.

If you have been using Word 2007, then Word 2010 will not seem that different. However, if you are a Mac user, the Ribbon at the top of the screen is a new addition to the user interface. The **Ribbon** is a single strip that displays **commands** in task-oriented **groups** on a series of **tabs** (Figure A1.1). In Word 2010, the new **File** tab replaces the Microsoft **Office Button** that was located in the top left of the screen in Word 2007. Additional commands in some of the groups can be accessed with the **Dialog Box Launcher**, a diagonal arrow in the right corner of the group label. The **Quick Access Toolbar** comes with buttons for saving your file and undoing and redoing commands; you can also add buttons for tasks you perform frequently.

The window size and screen resolution affect what you see on the Ribbon. You may see fewer command buttons or an entire group abbreviated to one button on a smaller screen or if you have set your screen display to a lower resolution. If you want to see the tab names without the command buttons, click **Minimize the Ribbon**.

The **Status Bar** at the bottom of the screen displays information about your document (Figure A1.2). Clicking the **Page position** button opens the **Go To** dialog box, which allows you to reposition the insertion pointer (the blinking vertical bar) on a different page or on a comment, table, graphic, or other landmark in your document. To return to your original position, click the **Undo** button on the **Quick Access Toolbar** at the top of the screen.

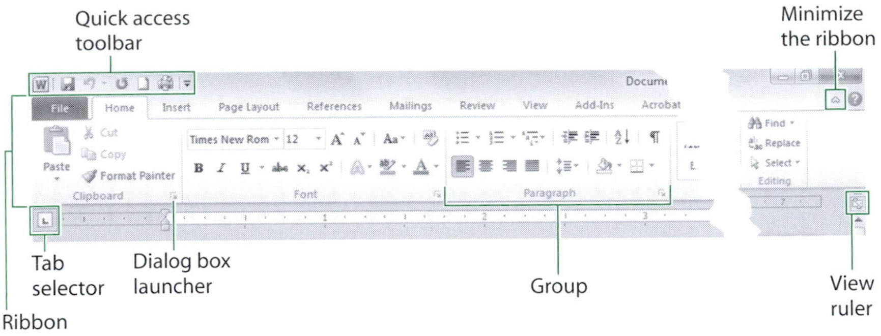

Quick access
toolbar

Minimize
the ribbon

Tab
selector

Dialog box
launcher

Group

View
ruler

Ribbon

Figure A1.1   Screen display of command area in a Word 2010 document.

The **Automatic word count** button updates the number of words as you type. The **Proofing** button shows a green check mark if Word has not found any spelling or grammatical errors; if it has, the button displays a red **X**. To correct the potential error, click the **Proofing** button and select one of the options in the dialog box.

To select which features are displayed on the status bar, right-click it to open the **Customize Status Bar** dialog box. Click the **Language** check box to display the language you are currently working in. If you have no occasion to use languages other than English, it is not necessary to display the **Language** button on the status bar.

The **Layout selector**, **Zoom level**, and **Zoom slider** buttons on the right side of the status bar are used to adjust how you view your document on the screen. They do not affect formatting.

Word 2010 has many features that make the design more attractive. There are more templates for letters, faxes, resumes, greeting cards, and more clipart and artwork than in earlier versions. The new default typefaces, Calibri

Page
position

Proofing

Layout
selector

Zoom slider

Page: 1 of 9   Words: 101   English (U.S.)   100%

Automatic
word count

Language

Zoom level

Figure A1.2   **Status bar** at the bottom of the screen. Clicking a button on the left side of the bar opens a dialog box or information window. The right side of the bar is used to change the view.

TABLE A1.1 Navigate document using keyboard shortcuts	
Where to?	**Hot Key**[a]
Beginning of the document	Ctrl+Home
End of the document	Ctrl+End
Beginning of a line	Home
End of a line	End
Up one screen	Page Up
Down one screen	Page Down
Up one page	Ctrl+Page Up
Down one page	Ctrl+Page Down

[a]The nomenclature Ctrl+Home means to hold down **Ctrl** while pressing **Home**.

and Cambria, were chosen for their improved on-screen appearance and versatility in web-based and printed media. Although the default font is smaller (11 pt instead of 12 pt), the line spacing is larger (1.15 instead of 1).

While Word 2010 can make your documents look pretty, it cannot teach you how to write scientific papers with proper content and style. That skill is something you can hope to acquire with practice. Consult this appendix to become more efficient at word processing: look up what you don't know and figure out how to do things faster. Ultimately you will be able to devote more time and energy to the content of your scientific papers than the format.

Increasing Your Word Processing Efficiency

Navigation

Use the keyboard shortcuts in Table A1.1 to move around the document quickly.

Navigation pane

The navigation pane is a convenient way to jump to different sections in longer documents. To display the navigation pane, click the **View** tab and check the **Navigation Pane** box in the **Show** group. The navigation pane has 3 tabs, which correspond to the Heading, Page, and Search Results views (Figure A1.3A). In **Heading view**, click a heading to move to that point in the document. Rearrange entire sections by dragging a heading to a new

(A)   (B)

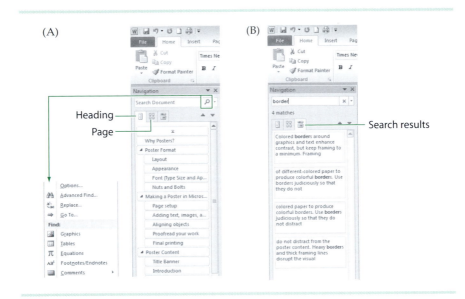

Heading
Page
Search results

Figure A1.3   The Navigation pane facilitates rapid navigation through long documents. (A) Click **Heading view** to jump to a particular section or **Page view** to look at a thumbnail of a particular page. (B) Click **Search results view** to go to each location where a word or another document element was found. Clicking the magnifying glass next to the search box provides more options.

position on the navigation pane. Similarly, clicking a page thumbnail in **Page view** allows you to move to that page. The third tab lets you search the document not just for words, but also graphics, equations, and comments (click the magnifying glass next to the search box to display options). The search results are displayed in boxes that, if clicked, move you to that location in the document (Figure A1.3B).

Text selection

Word offers you multiple ways to select text with the mouse or the keyboard or a combination of both. Word processing experts tend to prefer the keyboard because taking their hands off the keyboard to edit or format text with the mouse takes time. On the other hand, clicking command buttons with the mouse is more intuitive and does not require memorizing keyboard shortcuts. Table A1.2 shows you how to select text with both keyboard and mouse so that you can copy, move, format, or delete it. Before you follow these instructions, position the insertion pointer (the blinking vertical bar) next to the character, word or block of text you want to select.

TABLE A1.2	Select text with the keyboard and mouse after positioning the insertion pointer	
Text	**Keyboard**[a]	**Mouse**
Character	Shift + → or Shift + ←	Press and hold the left mouse button and drag over the character.
Word	Ctrl + Shift + → or Ctrl + Shift + ←	Double-click the word.
Sentence	Repeat sequence for a word	Hold down the **Ctrl** key and click.
Line	Shift + ↓ or Shift + ↑	Position the insertion pointer in the left margin so that it changes to an arrow. Click.
Block of text	Shift + arrow keys, Home, End, Page Up, or Page Down	Position the insertion pointer at the beginning of the block. Hold down the **Shift** key. Click at the end of the block.
Paragraph	Ctrl + Shift + ↓ or Ctrl + Shift + ↑	Triple-click anywhere in the paragraph.
Entire document	Ctrl+A	Position the insertion pointer in the left margin so that it changes to an arrow. Triple-click.

[a]The nomenclature Shift + → means to hold down **Shift** while pressing →; similarly Ctrl+Shift + → means to hold down both **Ctrl** and **Shift** while pressing →.

Commands in Word 2010

The best way to get up to speed with Word 2010 is to learn where the commands you used most frequently in earlier versions of Word are located. Table A1.3 is designed to help you with this task. Frequently used commands are listed in alphabetical order in the first column, and the Word 2010 command sequence is given in the second column. The PC keyboard shortcuts for most of the commands are shown in the third column. When using the Alt commands, hold down the **Alt** key while pressing the first letter. Release the **Alt** key and type the next letter(s) in the sequence. If you are a Mac user,

Ctrl = Command (⌘)

Alt = Option (however, some Alt key sequences do not have a Mac equivalent)

TABLE A1.3 Common commands (listed alphabetically) and how to carry them out in Word 2010[a]

Command	Word 2010 Command Sequence	PC Hot Key
Editing		
AutoCorrect	File | Options | Proofing | AutoCorrect Options	Alt+FT
Comments	Review | Comments	Alt+R
Copy	Home | Clipboard | Copy	Ctrl+C or Alt+HC
Cut	Home | Clipboard | Cut	Ctrl+X or Alt+HX
Find	Home | Editing | Find	Ctrl+F or Alt+HFDF
Paste	Home | Clipboard | Paste	Ctrl+V or Alt+HV
Paste Special	Home | Clipboard | Paste | Paste Special	Ctrl+Alt+V or Alt+HVS
Redo last action	Quick Access Toolbar | Redo button ↻	Ctrl+Y
Replace	Home | Editing | Replace	Ctrl+H or Alt+HR
Spelling & Grammar Check	Review | Proofing | Spelling & Grammar	F7 or Alt+RS
Track changes	Review | Tracking | Track Changes	Ctrl+Shift+E or Alt+RG
Undo		
Last action	Quick Access Toolbar | Undo button ↺	Ctrl+Z
Several previous actions	Undo button drop-down menu	
File		
Close	**X** (on right side of Title bar)	Ctrl + W or Alt+F4 or Alt+FX (Mac: Cmd+Q)
Document properties	File | Info	Alt+FIQ
Open existing	File | Open	Ctrl+O or Alt+FO
Open new	File | New	Ctrl+N or Alt+FN

[a]Many of the Word 2010 commands can also be accessed by selecting text or an object and right-clicking. For Alt commands, hold down **Alt** while pressing the first letter key, which corresponds to a tab name ![tab names] . Then release **Alt** and type the next letter(s) in the sequence. F1, F4, F7, and F12 are function keys. The nomenclature Ctrl+C means to hold down **Ctrl** while pressing **C**; similarly Ctrl+Shift+E means to hold down both **Ctrl** and **Shift** while pressing **E**.

TABLE A1.3 *Continued*		
Command	**Word 2010 Command Sequence**	**PC Hot Key**
Print	File | Print Quick Access Toolbar 🖶	Ctrl+P or Alt+FP
Save	File | Save Quick Access Toolbar 💾	Ctrl+S or Alt+FS
Save as	File | Save As	F12 or Alt+FA
Formatting		
Align text Left/center/ right/justified	Home | Paragraph	Ctrl+L/E/R/J or Alt+ HAL/HAC/HAR/HAJ
Font face, size, style, color	Home | Font	Ctrl+D or Ctrl+Shift +P or Alt+HFN
Format Painter Copy formatting	Home | Clipboard | Format Painter	Ctrl+Shift+C or Alt+HFP
Apply formatting	Drag cursor (Paintbrush I-beam) over text to be formatted	Ctrl+Shift+V
Lists Bulleted	Home | Paragraph | Bullets	Alt+HU
Numbered	Home | Paragraph | Numbering	Alt+HN

Show/Hide
  Formatting
  symbols
	Home | Paragraph | Show/Hide¶	Ctrl+Shift+* or
		Alt+H8

Gridlines	View | Show/Hide | Gridlines	Alt+WG
Ruler	View | Show/Hide | Ruler	Alt+WR
Spacing	Home | Paragraph dialog box launcher | Indents and Spacing tab	Alt+HPG
Lines	Line Spacing: drop-down menu	
Paragraphs	Spacing (Before: and After:)	
Subscript	Home | Font | x_2	Ctrl+ = or Alt+H5
Superscript	Home | Font | x^2	Ctrl+Shift+= or Alt+H6
Symbols	Insert | Symbols | Symbol	Alt+NU
Tabs	Home | Paragraph | Tabs or Page Layout | Paragraph | Tabs	Alt+HPG

TABLE A1.3   *Continued*

Command	Word 2010 Command Sequence	PC Hot Key
Page Layout		
Headers & Footers	Insert|Header & Footer|Header/Footer	Alt+NH/NO
Margins	Page Layout | Page Setup | Margins	Alt+PM
Orientation	Page Layout | Page Setup | Orientation	Alt+PO
Columns	Page Layout | Page Setup | Columns	Alt+PJ
Page break	Page Layout | Page Setup | Breaks | Page	Ctrl+Enter or Alt+PBP
Page numbers	Insert | Header & Footer | Page Number	Alt+NNU
Paragraphs		
Dialog box launcher	Home | Paragraph or Page Layout | Paragraph	Alt+PPG
Hanging indent	Dialog box launcher | Indentation | Special: Hanging	Ctrl+T
Indent first line	Dialog box launcher | Indentation | Special: First line	Ctrl+M, then Ctrl+Shift+T
Indent whole paragraph	Page Layout | Paragraph | Indent	Ctrl+M or Alt+PIL/PIR
Keep with next (prevents un-wanted page breaks)	Dialog box launcher | Line and Page Breaks tab | Pagination | Keep with next	Alt+HPGP
Shapes		
Align	Shift+Click individual shapes. Drawing Tools Format | Arrange | Align	Alt+JAA
Display gridlines	Click a shape. Drawing Tools Format | Arrange | Align | View Gridlines	Alt+JAAS
Format	Right-click the shape. Format Shape	Alt+JO
Group	Shift+Click individual shapes. Drawing Tools Format | Arrange | Group | Group	Alt+JAGG
Insert	Insert | Illustrations | Shapes	Alt+NSH
Nudge to position precisely	Click shape, then Ctrl+arrow keys.	
Position freely	Click a shape. Drawing Tools Format | Arrange | Align | Grid Settings | Uncheck Snap objects to grid when the gridlines are not displayed	Alt+JAAG

TABLE A1.3 *Continued*		
Command	**Word 2010 Command Sequence**	**PC Hot Key**
Special Characters		
Equations	Insert | Symbols | Equation	Alt+NE
Hyphen, nonbreaking		Ctrl+Shift+-
Space, nonbreaking		Ctrl+Shift+ spacebar
Symbols	Insert | Symbols | Symbol	Alt+NU
Tab stops	Home | Paragraph dialog box launcher | Tabs	Alt+HPG
Table of Contents		
Generate	References | Table of Contents | TOC	Alt+ST
Mark headings to use as entries	Home | Styles | Heading 1 (2 or 3)	Alt+HL or Ctrl+Alt+1/2/3
Update	References | Table of Contents | Update Table	Alt+SU
Tables		
Insert	Insert | Tables | Table	Alt+NT
Format	Click inside table, then Table Tools Design and Table Tools Layout tabs	Alt+JT (Design) Alt+JL (Layout)
Repeat table headings on multiple-page tables	Click anywhere in the first row, then Table Tools Layout | Data | Repeat Header Rows	Alt+JLJ
View		
Document views	View | Document | Views Status bar icons lower right	Alt+W
Multiple open documents side by side	View | Window | View Side by Side	Alt+WB
Multiple open documents horizontally	View | Window | Arrange All	Alt+WA
Print preview	File | Print	Alt+FP
Split current window (keep one part of document fixed while scrolling through the rest)	View | Window | Split	Alt+WS

In this book, the nomenclature for the command sequence is as follows: **Ribbon tab | Group | Command button | Additional Commands** (if available). For example, **Home | Editing | Find | Find** means "Select the **Home** tab on the Ribbon, and in the **Editing** group, click the down arrow on the **Find** button to select **Find** from the menu."

For an overview of what's new in Word 2010, go to Microsoft's Home Page (http://office.microsoft.com) and type "what's new in word 2010" into the search box. The link for Word for Mac 2011 is http://www.microsoft.com/mac/word/whats-new.

Microsoft Office also has an online interactive guide to help you transition from Word 2003 to Word 2010: Go to Microsoft's Home Page (http://office.microsoft.com) and type "office 2010 interactive" into the search box. Select "Learn where menu and toolbar commands are in Office 2010 and related products" and follow the instructions on the website.

Unfurling the Ribbon

If you prefer a more visual approach to the commands, the following sections show you the buttons on each tab. If you want more choices, click the dialog box launcher (the arrow on the right side of some groups) to open a dialog box for that particular task. The dialog boxes are mostly the same as those in Word 2003 and 2007.

The File tab

Clicking the **File** tab opens **Backstage View** (Figure A1.4). The left panel lists the file-related commands, the center panel provides the options for that command, and the right panel shows a preview or other options. Document commands such as Save, Save As, Open, Close, Recent, New, and Print, which were located on the **Office Button** in Word 2007 and on the **File** menu in Word 2003, are located on the **File** tab. Other commands on the **File** tab are

- **Info**, which provides new file management tools for setting permissions, inspecting the document for hidden properties and personal information, and recovering earlier autosaved or unsaved versions of a file.

- **Save & Send** makes it easier to collaborate with others on files. All collaborators must use Office 2010 and possess SharePoint 2010 software or a SkyDrive account to access and save the files online. Similar to Google docs, this feature allows two or more people to work on the same document at the same time.

(A) (B)

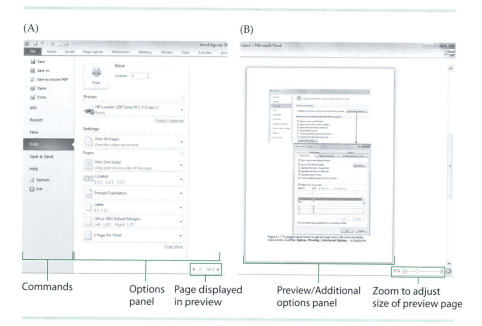

| Commands | Options panel | Page displayed in preview | Preview/Additional options panel | Zoom to adjust size of preview page |

Figure A1.4 Backstage View is displayed when clicking the **File** tab. (A) The left panel shows file-related commands such as Save, Save As, Open, Close, and Print. The options panel shows additional commands related to the command selected on the left panel, in this case **Print**. (B) The preview panel appears on the right in Backstage View and here shows page 2 as it will be printed. The zoom slider affects the view, but not the page format.

- **Help**. The **Help** menu can also be accessed by pressing the **F1** function key.
- **Options** allows you to change program settings such as user name and initials, rules for checking spelling and grammar, AutoCorrect options, how often documents are saved automatically, and so on. The **Options** button is also the gateway to commands for customizing the Ribbon and the Quick Access Toolbar and for managing Microsoft Office add-ins and security settings.

AutoCorrect. AutoCorrect is useful for more than just correcting spelling mistakes. Increase your efficiency by programming AutoCorrect to replace an expression that takes a long time to type with a simple keystroke combination. You must choose the simple keystroke combination judiciously, however, because every time you type these keystrokes, Word will replace them with what you programmed in AutoCorrect. One "trick" is to precede the keystroke combination with either a comma or semicolon (, or ;). Since punctuation marks are typically followed by a space and not a letter, adding one before your "code letter(s)" creates a unique combination.

For example, let's say "beta-galactosidase" is a word you have to type frequently. You choose ";bg" to designate "beta-galactosidase." To program AutoCorrect for this entry:

1. Select **File | Options | Proofing | AutoCorrect Options**. This opens the **AutoCorrect** tab in the **AutoCorrect** dialog box (Figure A1.5).
2. Type ";bg" in the **Replace** text box.
3. Type "beta-galactosidase" in the **With** text box.
4. Click the **Add** button.
5. Click **OK** to exit the **AutoCorrect** dialog box.
6. Click **OK** to exit the **Word Options** box.

AutoCorrect can also take the work out of formatting expressions with sub- or superscripts. Follow these steps:

1. First type the expression without sub- or superscripts (e.g., Vmax).
2. Select the characters to be sub- or superscripted, and then format them by clicking the appropriate button on **Home | Font** (the expression then becomes V_{max}).
3. Select the entire expression and click **File | Options | Proofing | AutoCorrect Options** (see Figure A1.5).
4. On the **AutoCorrect** tab, you will notice "Vmax" already entered in the **With** text box. Click the **Formatted text** option button to add the subscripting. The expression then becomes V_{max}.
5. Type "Vmax" in the **Replace** text box.
6. Click the **Add** button.
7. Click **OK** to exit the **AutoCorrect** dialog box.
8. Click **OK** to exit the **Word Options** box.

AutoCorrect can also be programmed to italicize scientific names of organisms automatically. By convention, scientific names of organisms are always italicized. When you type the name of the organism, capitalize genus but write species in lower case (e.g., *Homo sapiens*). To automate italicization, follow these steps:

1. First type the scientific name of the organism in your text (e.g., Aphanizomenon flos-aquae).

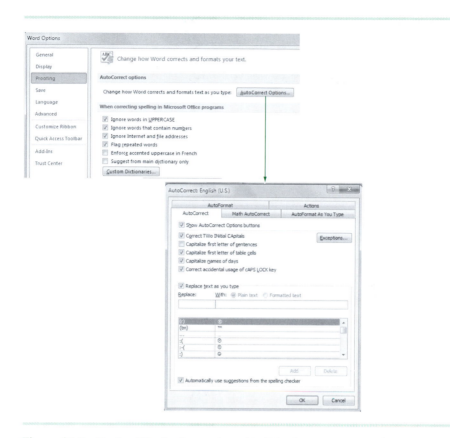

Figure A1.5   To simplify the formatting of italicized expressions or those with mathematical symbols, sub- and superscripts, program AutoCorrect. Click **File | Proofing | AutoCorrect Options** to open the dialog box.

2. Italicize the name by selecting it and then clicking the **Italic** button (**I**) on **Home | Font**. The text then becomes: *Aphanizomenon flos-aquae.*

3. With the name still selected, click **File | Options | Proofing | AutoCorrect Options** (see Figure A1.5).

4. On the **AutoCorrect** tab, you will notice Aphanizomenon flos-aquae (without italics) already entered in the **With** text box. Click the **Formatted text** option button to italicize it.

5. Enter a shorter version of the name in the **Replace** text box, for example "aphan".

6. Click the **Add** button.

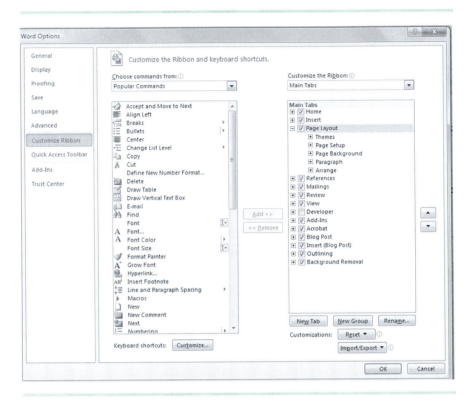

Figure A1.6   To add additional tabs with custom commands to the Ribbon, click **File | Options | Customize Ribbon** and then click **New Tab**.

7. Click **OK** to exit the **AutoCorrect** dialog box.

8. Click **OK** to exit the **Word Options** box.

Customize the Ribbon. If you frequently use commands that are buried deep within menus and sub-menus, then this feature is for you. Make custom tabs on the Ribbon, and add the hard-to-find commands for easy access. In Word 2010, click **File | Options | Customize Ribbon**. After clicking the **New Tab** button, add groups and commands to the new tab (Figure A1.6). It is not possible to customize the Ribbon in Word for Mac 2011, but the **menu bar** and the **standard toolbar** can be customized.

The Home tab

The **Home** tab contains commands for formatting text and paragraphs (Figure A1.7). The familiar **Find** and **Replace** buttons are also located here.

Paste
special

Show/Hide ¶

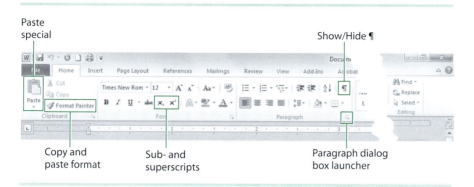

Copy and
paste format

Sub- and
superscripts

Paragraph dialog
box launcher

Figure A1.7   The **Home** tab highlighting selected command buttons.

Paste Special. Use Paste Special rather than Paste when you want to copy Web page content without the HTML (Hypertext Markup Language) formatting. Click the **Clipboard | Paste** down arrow and select **Paste Special | Unformatted Text** and then **OK**.

Format Painter. This paintbrush is great for copying and pasting format, whether it is the typeface or font size of a character or word, the indentation of a paragraph, or the position of tab stops. It's also handy for resuming numbering in a numbered list. It works with character, word, and paragraph formatting, but not page formatting.

To copy the format of a word once, select the word, click **Home | Clipboard | Format Painter** and drag the cursor (which has changed to a paintbrush next to an I-beam) over the text you want to format. To copy paragraph formatting (including tab stops), click inside the paragraph, but don't select anything. Then click the **Format Painter** and drag the mouse over the paragraph to which you want to apply the format.

Copy or move using drag and drop. If you have been using Word since middle school, you know how to copy and paste and cut and paste using the icons in the **Clipboard** group. You probably use the keyboard shortcuts Ctrl+C, Ctrl+X, and Ctrl+V and the right mouse button as well. A fourth way to copy or move text is Drag and Drop. This method is best for short distances within view of your screen; it works within and between documents (e.g., when you have two or more documents open side by side).

To *move* text with Drag and Drop, select the block (see Table A1.2) and hold down the left mouse button. The arrow changes to an arrow with a box. Continue to hold down the mouse button and drag the selected text to a new insertion point (location).

To *copy* text with Drag and Drop, select the block and hold down the **Ctrl** key while holding down the left mouse button. The arrow changes to an arrow with a plus sign in the box. Drag the selected text to a new insertion point.

If you expect to paste the same block more than once, use the Clipboard icons or keyboard shortcuts rather than Drag and Drop. Drag and Drop does not copy to Clipboard.

Sub- and superscripts. Writers and journalists may wonder why these buttons have been placed in such a prominent location in Word 2010. But having them there saves scientists quite a few keystrokes! To superscript or subscript text, first type the expression without formatting. Then highlight the character(s) to be sub- or superscripted and click the appropriate button in the **Font** group. Alternatively, you can apply and remove sub- and superscripting using keyboard commands ("hot keys"). Select the character(s), and then hold down **Ctrl** while pressing = for subscript or hold down both **Ctrl** and **Shift** while pressing = for superscript.

Expressions with sub- or superscripting are common in the natural sciences and their formatting makes them readily identifiable to members of the scientific community. Therefore, it would be incorrect to write sub- or superscripted characters on the same line as the rest of the text. Similarly, when using scientific notation, exponents are always superscripted. It is not acceptable to present exponents preceded by a caret (^) or by an uppercase E to designate superscript.

RIGHT: 2×10^{-3} (exponent is superscripted)

WRONG: $2 \times 10{-}3$ or $2 \times 10^{-}3$ or $2 \times 10E{-}3$

If you frequently have to type expressions with sub- and superscripts, save time by programming them in AutoCorrect (see p. 174).

Show/Hide. This button allows you to view formatting marks such as ¶ (paragraph), → (tab), and · (space). Normally you would not display these marks, because they are distracting. You might *want* to see them, however, when you are troubleshooting formatting problems.

Paragraph dialog box. The **Paragraph** dialog box can be accessed from both the **Home** and **Page Layout** tabs. Clicking the diagonal arrow on the **Paragraph** group label launches a dialog box (Figure A1.8) where you can set

- **First line indent** to denote the beginning of a new paragraph. Next to **Indentation | Special**, select **First line** from the drop-down menu and then specify **By: 0.5"** (or your preference).

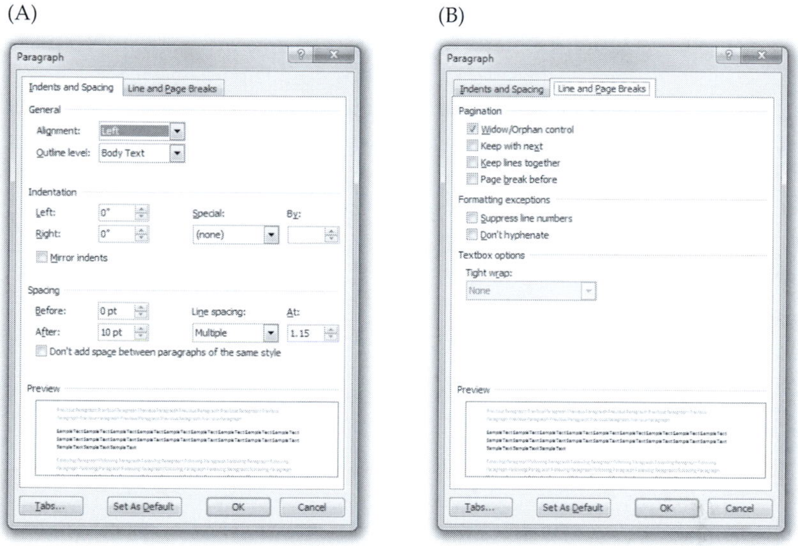

(A)

(B)

Figure A1.8   The **Paragraph** dialog box is launched by clicking the arrow on the **Paragraph** group label. (A) Click the **Indents and Spacing** tab to format first line and hanging indents (under "Special"); to adjust spacing between lines and between paragraphs; and to set tab stops. (B) Click the **Line and Page Breaks** tab and check "Keep with next" to prevent section headings from separating from the body.

- **Hanging indent** for listing full references. The first line of each reference begins on the left margin and the subsequent lines are indented. First type each reference so that it is aligned on the left margin and ends with a hard return ¶. When you are finished, select all of the references (see Table A1.2, Block of text) and click **Home | Paragraph** dialog box launcher. Next to **Indentation | Special**, select **Hanging** from the drop-down menu and then specify **By: 0.25"** (or your preference).

- **Line spacing** for all lines of text. The default spacing in Word 2010 is **Multiple at 1.15 lines**; your instructor may ask you to change the spacing to **Double** in order to have a little extra space to write comments. There is also a **Line spacing** command button in the **Paragraph** group on the Ribbon where you can make this change.

- **Paragraph spacing.** Besides indenting the first line, paragraphs can be separated with a blank line. To automatically add extra space after each paragraph, thereby saving you the trouble of pressing the **Enter** key twice, put the insertion pointer any-where in the paragraph. In the **Paragraph** dialog box, select **12 pt** in the **Spacing | After** box. This spacing corresponds to one blank line when using a 12 pt font size for your paper.

- **Line and page breaks.** This feature is handy for preventing section headings from being separated from the body due to a natural page break. Select the section heading (for example, "Materials and Methods"). Then open the **Paragraph** dialog box and click the second tab called **Line and Page Breaks**. Under **Pagination**, check the box next to **Keep with next** to keep the heading together with the body. To keep all the lines of a para-graph together on the same page, check **Keep lines together**.

- **Tabs.** Whenever you want to align text in columns, use tab stops or make a table without borders. Never use the space bar for more than one space. Even the old custom of putting two spaces after periods and colons has become obsolete as a result of proportionally spaced computer fonts. The **Tabs** button is located in the bottom left corner of the **Paragraph** dialog box. To add a new tab stop, open the **Tabs** dialog box and enter a **Tab stop position**. Choose an alignment (left, center, right, or decimal). Click the **Set** button. Repeat this procedure to add more tab stops. When you are done, click **OK**. The new tab will appear on the ruler at the top of the page. If the ruler is not dis-played, click **View Ruler** at the top of the scroll bar on the right side of the page (see Figure A1.1).

A faster, but slightly less precise, way to add tab stops is to select an align-ment by repeatedly clicking the **Tab selector** button on the left side of the ruler (see Figure A1.1) and then simply clicking the ruler where you want to add the new tab stop.

To remove the tab stop, point to it, hold down the left mouse button, and drag it off the ruler. To move the tab stop, simply drag it to another position on the ruler. You can also carry out these tasks in the **Tabs** dialog box.

The default units of the ruler are inches. If you would like to use metric units instead, click **File | Options | Advanced**. In the **Display** area, next to **Show measurements in units of**, select the desired units from the drop-down menu.

The Insert tab

The **Insert** tab has buttons for inserting pages and page breaks, tables, illus-trations, links, headers & footers, text boxes, equations, and symbols (Fig-ure A1.9).

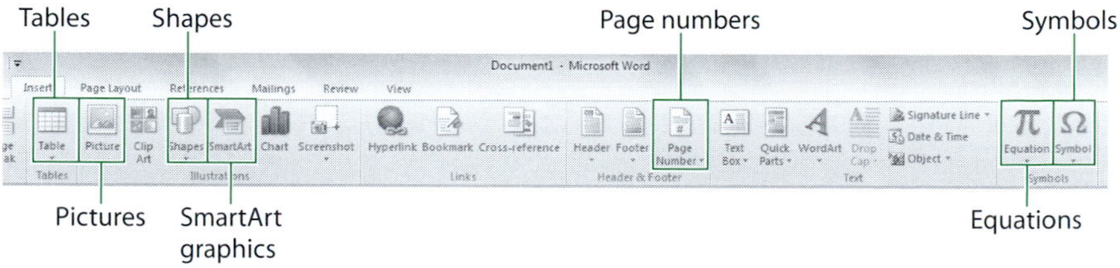

Tables   Shapes   Page numbers   Symbols

Pictures   SmartArt   Equations
            graphics

Figure A1.9   The **Insert** tab highlighting selected commands.

Tables. By convention, tables in scientific papers do not have vertical lines to separate the columns, and horizontal lines are used only to separate the table caption from the column headings, the headings from the data, and the data from any footnotes (notice the format of the tables in this book).

CREATING A TABLE

1. Position the insertion pointer where you want to insert the table.
2. Click **Insert | Tables | Table**. Highlight the desired number of columns and rows.

A blank table appears with the cursor in the first cell. To change the table format, use the **Table Tools Design** and **Layout** tabs (Figure A1.10). These tabs only appear when the insertion pointer is inside the table.

3. To insert or delete columns or rows or to merge or split cells, use the command buttons on the **Table Tools Layout** tab.

FORMATTING TEXT WITHIN TABLES

1. To apply formatting to adjacent cells, first select the cells by clicking the first cell in the range, holding down the **Shift** key, and then clicking the last cell.
2. To apply formatting to selected cells, repeat (1) using the **Ctrl** key.

FORMATTING A TABLE BY MODIFYING A BUILT-IN STYLE

1. Select all cells in the table and click **Design | Table Styles**. Click one of the "Light Shading" styles. These table styles have the appropriate horizontal lines and no vertical lines preferred in scientific papers.

(A)

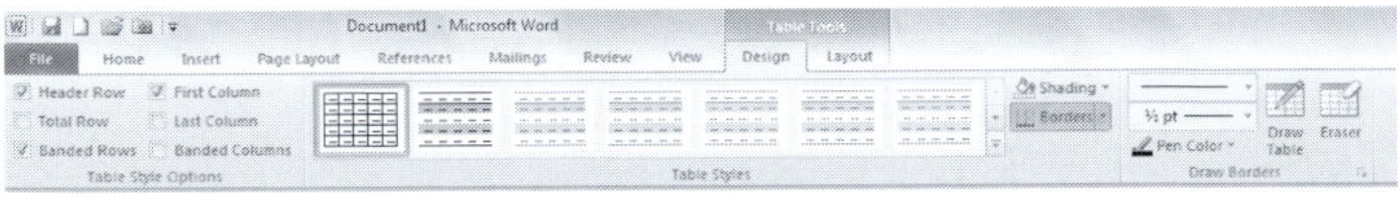

(B)

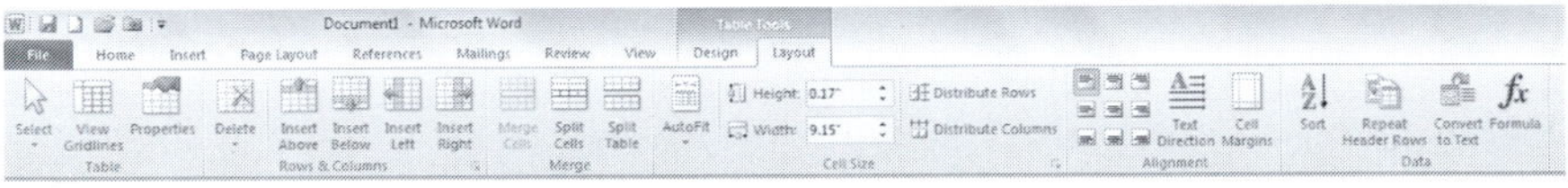

Figure A1.10 Table formatting tabs appear when you click inside a table. (A) The **Design** tab is used to modify shading and borders. (B) The **Layout** tab is used to insert or delete rows and columns, split or merge cells, adjust cell size, and sort data.

2. With the whole table still selected, click the down arrow next to the **Shading** button and select **No Color**.

NAVIGATING IN TABLES

- To jump from one cell to an adjacent one, use the arrow keys.
- To move forward across the row, use the **Tab** key. Note: If you press **Tab** when you are in the last cell of the table, Word adds another row to the table.
- To align text on a tab stop in a table cell, press **Ctrl+Tab**.

VIEWING GRIDLINES

It is easier to enter data in a table when the gridlines are displayed. Position the insertion pointer inside the table and click **Layout | Table | View Gridlines**. Gridlines are not printed.

ANCHORING IMAGES WITHIN TABLES

Getting images to line up horizontally and vertically on a page can be tricky. It's possible to make multiple columns on the **Page Layout** tab or to adjust the position of each picture on the **Picture Tools Format** tab, but quite often the images become misaligned when the document is edited. To make images stay put, create a table. Select the desired layout (for example, 3 columns × 1 row), size the images so they fit into the cells, and then paste the images into the table.

REPEATING HEADER ROWS

When a table extends over multiple pages, it is convenient to have Word automatically repeat the header row at the top of each page. To activate this command, click anywhere in the header row and then select **Table Tools Layout | Data | Repeat Header Rows**.

Picture. To insert a picture, click **Insert | Illustrations | Picture**. While that function is not new, the ability to edit the picture in Word is. When you click the picture, the **Picture Tools Format** tab appears. The buttons on this tab make it possible to remove the background, adjust brightness and contrast, apply artistic effects, and add various borders. Because high-resolution images can increase the size of a Word document significantly, compress the pictures if the document will be printed, posted to the Web, or shared via email (**Picture Tools Format | Adjust | Compress Pictures**).

Shapes. To make line drawings or add simple graphics to your document, click **Insert | Illustrations | Shapes**. To insert a line, for example, click a line to display crosshairs. Position the mouse pointer on the page where you want the line to start, hold down the left mouse button, drag to where you want the line to end, and release the mouse button. To make perfectly horizontal or vertical lines, hold down the **Shift** key while holding down the left mouse button and drag. In general, when it's important that a shape retain perfect proportions (e.g., circle not oval; square not rectangle), hold down the **Shift** key while drawing or resizing the shape with the mouse.

To change the angle or length of the line, click the line and then position the mouse pointer over one of the end points of the line to display a two-headed arrow. Hold down the left mouse button, drag to where you want the line to end, and release the mouse button.

To change the location of the line, click the line and then position the mouse pointer over the line to display crossed arrows. Hold down the left mouse button, drag the line to the desired location, and release the mouse button. For more precise positioning, right-click the line and select **Format AutoShape** from the drop-down menu. On the **Layout** tab, click the **Advanced** button. Change the **Picture Position** to place the line exactly where you want it.

To format the line, right-click it and select **Format AutoShape**. This dialog box gives you options for changing the appearance of the line, resizing it precisely, and laying it out in relation to other objects. Many of the same formatting commands are located on the **Drawing Tools Format** tab that appears when you select the line (Figure A1.11).

If several objects make up a graphic, then it may make sense to group them as a unit. Grouping allows you to copy, move, or format all the objects in the graphic at one time. To group individual objects into one unit, click

Format fill, line color and style,
and add special effects

Figure A1.11 To format a shape, click it to open the **Drawing Tools Format** tab. Right-clicking the shape or selecting the dialog box launcher opens a dialog box with additional options.

each object while holding down the **Shift** key. Release the **Shift** key and then click **Group** under **Drawing Tools Format | Arrange**. One set of selection handles surrounds the entire unit when the individual objects are grouped. To ungroup, simply select the group and click **Group | Ungroup**.

The **Align** button is handy for precisely lining up objects (shapes as well as text boxes). Click each object that you want to line up while holding down the **Shift** key. Release the **Shift** key, click **Align** under **Drawing Tools Format | Arrange**, and select one of the alignment options. Another useful feature for arranging objects on a page is **Align | View Gridlines**. This command puts a non-printing grid on your page. Click **Align | Grid Settings** to adjust the spacing between the gridlines or to snap objects to the grid. Check the **Snap Objects to grid when the gridlines are not displayed** box if you want to position objects on gridlines. Do not check this box if you prefer to position the objects freely. Finally, to move objects just a fraction of a millimeter from their current position, click the object, hold down the **Ctrl** key, and use the arrow keys to nudge the object exactly where you want it on the page.

SmartArt. SmartArt is a collection of MS Word graphics templates for lists, processes, cycles, and relationships. Before you go to the trouble of drawing your own graphics, see if one of the SmartArt templates might suit your needs. If you have to draw more complicated graphics, try Microsoft Visio. For highly technical drawings, a computer-aided design program like Auto-CAD®, SolidWorks, or Pro/ENGINEER, among others, may be required.

Page numbers. Page numbers make it easier for you to assemble the pages of your document in the correct order. To insert a page number, click **Insert | Header & Footer | Page Number** (see Figure A1.9) and then select a position for the numbers. As shown in Figure A1.12, after the page number is inserted at the top right of the page, the **Header & Footer Tools Design** tab appears. In the **Options** group, you have the opportunity to check

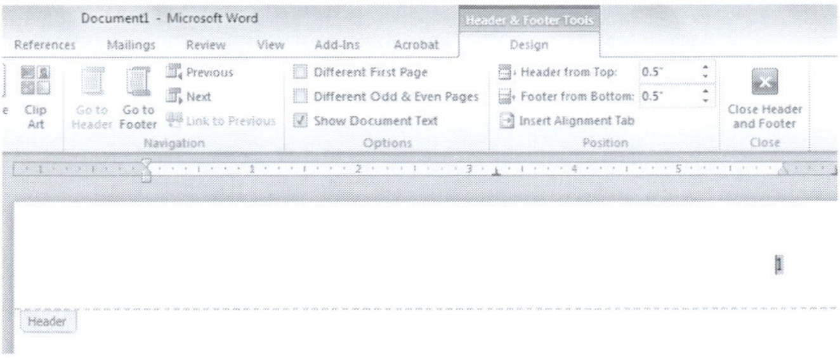

Figure A1.12   Page numbers are part of headers and footers. To remove the page number from the top of the first page, select **Different First Page** from the **Header & Footer Tools Design** tab.

Different First Page to remove the page number from the first page. To remove all page numbers, click **Insert | Header & Footer | Page Number | Remove Page Numbers**. This command is not available in documents saved in earlier versions of Word. For such documents, double-click the page number to open the header or footer and then double-click the page number again and press **Delete**. When you are finished inserting and formatting the page numbers, click **Close Header and Footer**.

Equations. To insert a common mathematical equation or make your own, click **Insert | Equation** to display the **Equation Tools Design** tab (Figure A1.13A). Peruse the built-in equations in the drop-down menu for the **Equation** button on the left side of the tab. If this menu does not contain the equation you'd like to write, choose a form to modify from the **Structures** group. Fill in numbers or select Greek letters, arrows, and mathematical operators from the 8 symbol sets that appear after clicking the **More** arrow (Figure A1.13B).

Symbols. Greek letters and mathematical symbols can also be inserted in running text. To do so, click **Insert | Symbol | More Symbols** (Figure A1.14A). The characters that are displayed in the **Symbol** dialog box depend on which font is selected. Under **Font: (normal text)**, you'll find commonly used scientific symbols such as degrees (°), plus or minus (±), micro (µ), parts per thousand (‰), infinity (∞), is less than or equal to (≤), is greater than or equal to (≥), middle dot (·), arrows for chemical reactions (← and →), female ♀, and male ♂. Normal text font also includes Greek letters and characters with diacritical marks used in European languages other than English (for example, à, â, ã, ä, and ç).

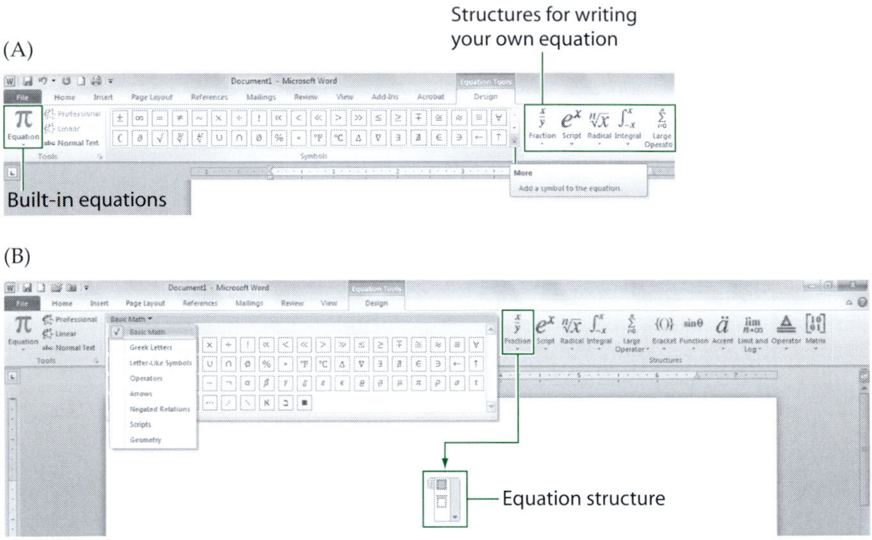

(A)

Structures for writing
your own equation

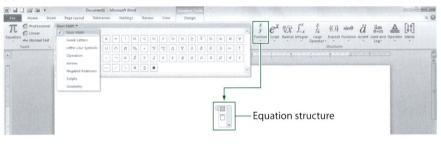

Built-in equations

(B)

Equation structure

Figure A1.13 (A) The **Equation Tools Design** tab lets you choose built-in equations or select a structure to modify. (B) Insert symbols, Greek letters, and operators for the equation by clicking the **More** arrow in the **Symbols** group and selecting from one of the symbols sets.

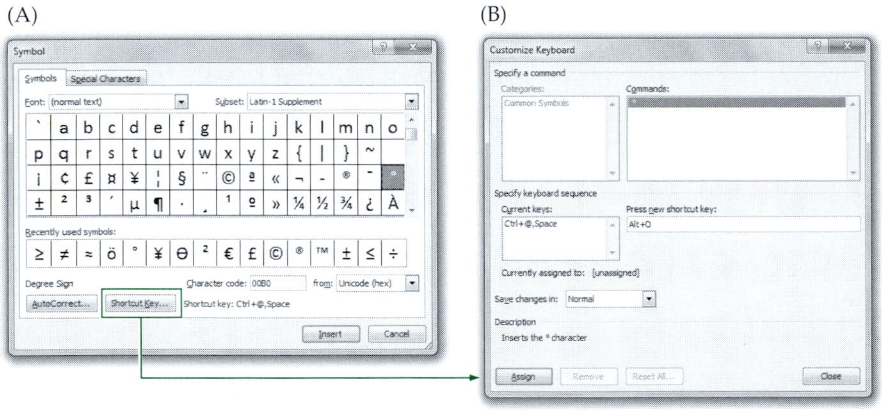

(A)

(B)

Figure A1.14 (A) The **Symbol** dialog box showing characters under **Font: (normal text)**. (B) **Customize Keyboard** dialog box allows you to define a keystroke combination for frequently used symbols.

If you use a particular Greek letter or mathematical symbol frequently, you can make a shortcut key to save time. Let's use the degree sign as an example (see Figure A1.14A). After clicking this symbol in the **Symbol** dialog box, click the **Shortcut Key** button at the bottom of the box. In the **Customize Keyboard** dialog box (Figure A1.14B), define a combination of keystrokes using Ctrl, Alt, Ctrl+Shift, or Ctrl+Alt plus some other character. Because the degree sign looks like a lower case *o*, let's use Alt+o. To define Alt+o as the shortcut key for the degree sign, type Alt+o in the **Press new shortcut key** box. Word notifies you that this combination is unassigned (as in Figure A1.14B) or that it has already been assigned to another command. (Note: Even though you typed Alt and lowercase o, the shortcut appears as Alt and uppercase O in the box. If you had typed Alt and uppercase O, this would have appeared as Alt+Shift+O.)

If the shortcut is already assigned to a different command, determine if you use that command very often. If you don't, you can remove the shortcut from the seldom-used command and assign it to the symbol you want. To do this:

1. Select **Assign** from the buttons at the bottom of the dialog box (see Figure A1.14B). Alt+O will then be listed under **Current keys**. If Word assigned a different shortcut key to the degree sign, this will also be displayed under **Current keys** (in this example, Ctrl+@, Space).

2. Click the previously assigned shortcut key, and then select **Remove** from the buttons at the bottom of the dialog box.

3. Close the **Customize Keyboard** dialog box.

4. Close the **Symbol** dialog box.

Next time you have to write the symbol for "degrees Celsius," simply type Alt+o followed by uppercase C to get: °C.

Another way to simplify inserting symbols is to program AutoCorrect. Once again using the degree sign as an example, open the **Symbol** dialog box, click the degree sign, and then click the **AutoCorrect...** button at the bottom of the box (see Figure A1.14A). The ° symbol will already be entered in the **With** text box and the **Formatted text** option box will be selected. Type a unique combination of characters in the **Replace** text box. Because you may not want to replace the word "degrees" with the degree symbol (°) every time, use the trick of preceding the word with either a comma or semicolon (, or ;) (see the "AutoCorrect" section on pp. 173–176). Type ";degrees" in the **Replace** box. Click the **Add** button, click **OK** to exit the **AutoCorrect** dialog box, and finally click **OK** to exit the **Word Options** box.

Figure A1.15   The **Page Layout** tab is used to format pages.

The Page Layout tab

The **Page Layout** tab has buttons for formatting whole pages (Figure A1.15). The **Page Setup** and **Paragraph** groups are the ones you're likely to use for your lab reports. The dialog box launchers for these groups open the dialog boxes that you are familiar with from using previous editions of Word. The commands in the **Themes** and **Page Background** groups affect the overall design of the page and are useful for designing your own stationery, invitations, and so on.

Document orientation.   Most lab reports are printed in Portrait Orientation. Occasionally, however, it may be necessary to insert a figure or large table in Landscape Orientation. To do so, insert a section break (*not* a page break) at the end of the previous page by clicking **Page Layout | Page Setup | Breaks | Section Breaks | Next Page**. Then click **Page Layout | Page Setup | Orientation | Landscape** to create a page turned sideways. After you are finished typing the table or pasting a figure created in Excel, insert another section break and choose Portrait to return to the usual orientation.

Sometimes section breaks mess up the page numbering. If this happens to your document, link the header or footer that contains the page number to the previous section. To do so, click **Insert | Header & Footer** and select either **Header** or **Footer**, depending on where the page number is located. Then click **Edit Header** (or **Edit Footer**) and scroll to the section in which the page number is messed up. Under **Header & Footer Tools Design | Navigation**, click **Link to Previous** to continue the numbering from the previous section. Click **Close Header and Footer** to exit (see Figure A1.12).

The References tab

Insert Endnotes.   The most useful feature on the References tab, at least with regard to scientific papers, is the Insert Endnote (*not* footnote!) command for in-text references in the Citation Sequence (C-S) system (Figure A1.16). To insert superscripted endnotes, type the sentence containing information that requires a reference. End the sentence with a period. Then click **References | Footnotes | Insert Endnote**. Type the full reference in proper for-

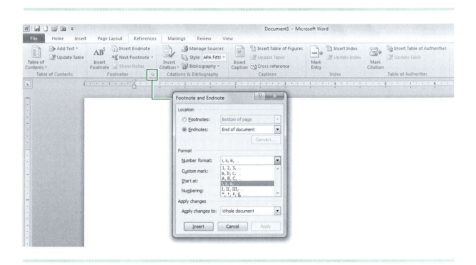

Figure A1.16   The **Footnote and Endnote** dialog box opens after clicking the **Footnotes** dialog box launcher on the **References** tab. Number format is selected from the drop-down menu.

mat (see "The Citation-Sequence System" in Chapter 4). Click **References | Footnotes | Show Notes** to exit the endnote and return to the text. To change the endnote number style from Roman to Arabic, click the **Footnotes** dialog box launcher and select Arabic numeral format from the drop-down menu.

Word updates endnotes automatically so that they appear sequentially. To see an endnote, point the mouse at the superscripted number. To edit an endnote, double-click the number. To delete an endnote, highlight the superscripted number and press **Delete**. The other endnote numbers will be renumbered automatically.

Table of contents. For longer papers, such as dissertations and honors theses, you might like to provide a table of contents. The easiest way to generate a table of contents is to use the built-in heading styles on the **Home** tab. First decide which headings (and the level of each) in your document will be used as entries in the table of contents. Second, generate the table.

MARK THE ENTRIES AS HEADINGS IN YOUR DOCUMENT

It doesn't matter if you define the heading level before or after you type the heading. To define the heading level first, click **Home | Styles** and then select **Heading 1 (2, 3, or 4)**. Then type the heading and press **Enter**. Alternatively, to mark the text as a heading after you've typed it, highlight it, and then click **Home | Styles** and select the appropriate heading level.

GENERATE THE TABLE OF CONTENTS

Place the insertion pointer at the beginning of the document. Click **References | Table of Contents | Table of Contents** and select a style for the table. Word automatically generates the table from the built-in headings. If you add or delete headings while revising your document, simply click **References | Table of Contents | Update Table**.

Citations & Bibliography. The **Citations & Bibliography** group promises to cite sources in one of the common styles, but this feature has major deficiencies. First, CSE style is not among those listed. Second, APA style (American Psychological Association) has similarities to CSE style, but the in-text citation comes out wrong. Whether the publication has one, two, or more authors, Word only lists one author in the in-text reference. The *correct* way to cite references in the Name-Year (N-Y) system is to give both authors' names when there are two authors and the first author's name plus *et al.* when there are three or more. Until Word's software developers are able to correct these mistakes, use another references management tool or type in-text references and full references as described in Chapter 4.

The Review tab

The **Review** tab is where you'll find commands for checking spelling and grammar, looking up synonyms and antonyms, adding comments, tracking changes made by others, and other functions related to editing (Figure A1.17).

Proofing. The Proofing group contains a spelling and grammar checker and buttons that connect to online dictionaries and other references. Spelling and grammar issues are discussed on pp. 193–194.

Comments. Comments are frequently used to exchange ideas when collaborating on a paper. You are most likely to add a comment, read someone else's comment and respond with a comment of your own, or delete one or more comments. You may also like to change how the comments are displayed on your screen.

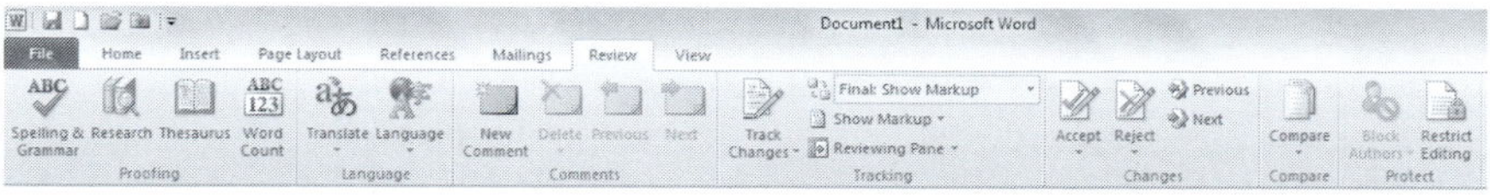

Figure A1.17   The **Review** tab provides options for revising your document, including spelling and grammar check, making comments, and tracking changes.

ADDING A COMMENT

1. Select the text you want to comment on.
2. Click **Review | Comments | New Comment**.
3. Type your comment in the box. Click outside of the comment area to return to the document.
4. To change the initials used in comments, go to **File | Options | General | Personalize your copy of Microsoft Office**.

REVIEWING COMMENTS

When there are no comments in a document, the **Delete**, **Previous**, and **Next** buttons in the **Comments** group are grayed out (as in Figure A1.17). When there are comments, these commands become available. After you have read and, if appropriate, taken action on all of the comments, delete them with **Review | Comments | Delete | Delete All Comments in Document**. To delete comments one at a time, right-click the balloon or the inline comment and select **Delete Comment**.

DISPLAYING COMMENTS

Comments can be displayed as balloons in the margin (as in Figure A1.18) or within the document itself (inline). To see the difference, click **Review | Tracking | Show Markup | Balloons** and click one of the three options: **Show Revisions in Balloons**, **Show All Revisions Inline**, or **Show Only Comments and Formatting in Balloons**. When the inline option is selected, revisions appear in a Reviewing Pane at the bottom or left-hand side of the screen (depending on the layout selected under **Tracking | Reviewing Pane**).

Tracking changes made by others.   It may not always be possible for you and your collaborator (or reviewer) to find a common time to meet to go over your paper. Email makes the review process more convenient. You can

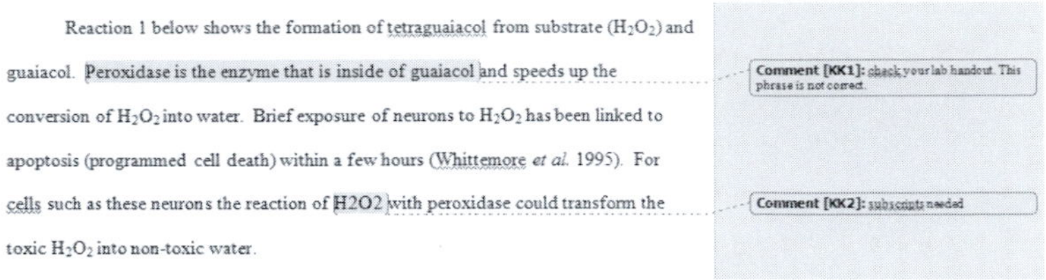

Reaction 1 below shows the formation of tetraguaiacol from substrate (H₂O₂) and

guaiacol. Peroxidase is the enzyme that is inside of guaiacol and speeds up the

conversion of H₂O₂ into water. Brief exposure of neurons to H₂O₂ has been linked to

apoptosis (programmed cell death) within a few hours (Whittemore *et al.* 1995). For

cells such as these neurons the reaction of H2O2 with peroxidase could transform the

toxic H₂O₂ into non-toxic water.

Comment [KK1]: check your lab handout. This phrase is not correct.

Comment [KK2]: subscripts needed

Figure A1.18   Comments are displayed in numerical order in balloons (as in this screenshot) or on a Reviewing Pane. Command buttons related to viewing, adding, and deleting comments are located on the **Review** tab.

send your paper to a reviewer in electronic format as an attached file, and your reviewer can send it back to you after making comments or revisions directly in the document.

It is important for you to be able to distinguish your original text from the comments and changes suggested by your reviewer. After all, you are the author, who has the right to accept or decline the reviewer's suggestions.

Before you send your lab report to a reviewer, click **Review | Tracking | Track Changes** to turn on tracking (the button will turn orange). As long as this command is activated, anything someone types in the document will be colored and underlined. Anything someone deletes will be colored and crossed out. Each reviewer gets his or her own color so you can identify the changes made by multiple reviewers. In addition, a black vertical line in the left margin alerts you to changes in your document.

When you get your document back, open it, and turn off tracking by clicking the **Track Changes** button again. Then click **Changes | Next** to go to the first location where a comment or revision was made. Use the drop-down menus for **Accept** or **Reject** or simply click **Next** to read the suggestion without taking any action.

To make sure you've addressed all tracked changes, save your document and then click **File | Info | Check for Issues | Inspect Document**. The **Document Inspector** box will notify you if any comments, revision marks, personal information, or other issues were found. Rather than clicking **Remove All**, close the box and look for the changes you missed. To do so, click **Review | Changes | Next**. Delete the stray revision mark after deciding whether any action is necessary.

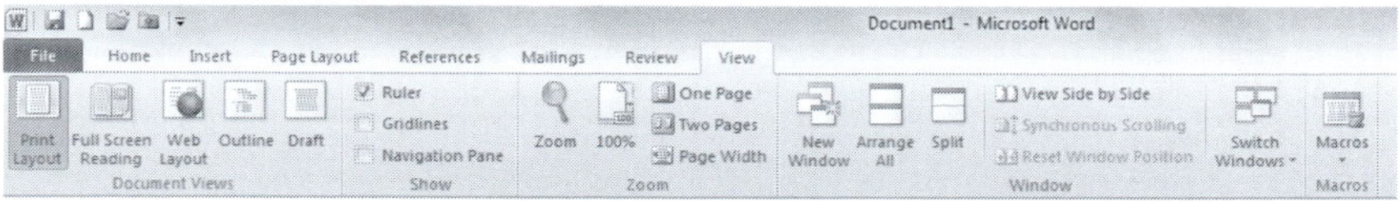

Figure A1.19   The **View** tab contains commands that were present on the **View** and **Window** buttons on the menu bar in Word 2003.

The View tab

Print Layout view is the one you see when you open a Word 2010 document. This view shows you how your page will look when it's printed, including headers and footers and images. Other views can be selected from the **Document Views** group on the **View** tab (Figure A1.19) and on the status bar at the bottom of the screen (see Figure A1.2, Layout selector).

Proofreading Your Documents

Before you waste reams of paper printing out drafts of your document, have Word check spelling and grammar and inspect the document for stray revision marks and comments. Use your eyes to look over format on-screen before proofreading the printed document.

Spelling and grammar

Word gives you visual indicators to alert you to possible spelling and grammar mistakes:

- The **Proofing** button on the status bar at the bottom of the screen shows a red *X* instead of a green ✓
- Words that are possibly misspelled are underlined with a wavy red line
- Phrases that may contain a grammatical error are underlined with a wavy green line.

Do not ignore these visual cues! Click the **Proofing** button to see what potential problem(s) Word has identified. Take action by choosing one of the options in the dialog box. Alternatively, to deal with a word underlined in red, right-click it. A pop-up menu appears with commands and suggestions for replacements (Figure A1.20). **Ignore All** applies only to the current document. **Add to Dictionary** applies to all future documents. It makes sense to add scientific terminology to Word's dictionary after you consult

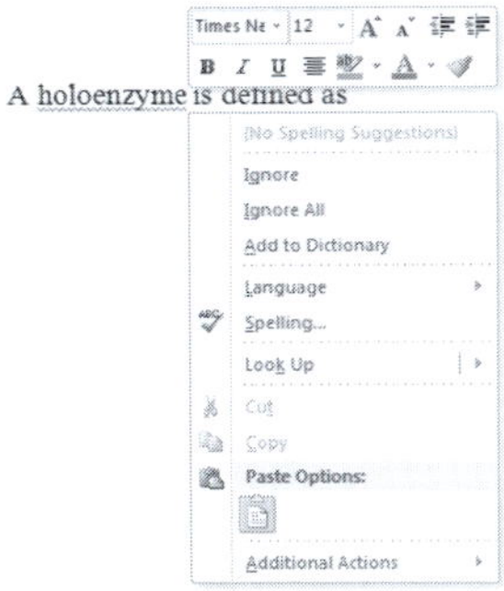

Figure A1.20   Dialog box for spelling suggestions and other ways to handle words not in Word's dictionary. When you right-click a word in Word 2010, a mini-toolbar also pops up with options for changing typeface, character size, color, and so on.

your textbook or laboratory manual to confirm the correct spelling. After making a selection on the pop-up menu, the wavy red underline is deleted and the word is ignored in the manual, systematic spelling and grammar check. Similarly, to deal with a possible grammatical error underlined in green (including extra spaces between words), right-click it to accept or ignore Word's suggestions.

The **AutoCorrect** function corrects common types of spelling mistakes as you type. To see the list of commonly misspelled words that Word corrects automatically, click **File | Options | Proofing | AutoCorrect Options**, and scroll through the words on the AutoCorrect tab. If you do not want Auto-Correct to change a particular keystroke combination (e.g., do not change (c) to ©), then select this combination on the list, and click the **Delete** button. You can also add to the list words that you know you misspell frequently.

If AutoCorrect changes text that you don't want changed while you are typing, click the **Undo** button on the **Quick Access Toolbar**.

Format

Use **Print Layout** view to proofread your document on screen. This view closely resembles what your printed page will look like with headers, footers, blank areas, figures, and page breaks. **Print Layout** is the view you see when you first open a Word 2010 document.

When proofreading for format, pay attention to the following items:

Margins. 1.25″ left and right and 1″ top and bottom margins give your instructor room to make comments. Margins can be adjusted by clicking **Page Layout | Page Setup | Margins** or **Page Layout | Page Setup** dialog box launcher.

Paragraph spacing. Paragraphs end with a ¶ symbol. The next paragraph starts either with the first line indented or preceded by an empty line. Word 2010 automatically adds 10 pt spacing after each paragraph to simplify formatting. If you like the automatic space, then do not use the first line indent method for starting a new paragraph. If you prefer the first line indent method, go to **Home | Paragraph** dialog box launcher and change **Indentation: Special** to **First line** and **Spacing: After** to **0**.

Document orientation. Most lab reports are printed in Portrait Orientation. If, however, you have one or more pages in Landscape Orientation to accommodate a figure or large table, for example, check the format of the margins, page numbers, and orientation of *all* the pages in the document. If you find any orientation-related formatting problems, correct them with commands on the **Page Layout** tab.

Page numbers. Page numbers make it easier for you to assemble the pages of your document in the correct order. Check that the pages are numbered consecutively, especially the page after a section break. See the "Page Numbers" section on pp. 184–185 to insert or remove page numbers. See "The Page Layout tab" on p. 188 for troubleshooting tips.

Section headings must not be separated from the body. Research articles are divided into sections: Abstract, Introduction, Materials and Methods, Results, Discussion, References, and Acknowledgments. Each section begins with a heading, on a separate line, followed by the body. When you check the format of your paper, make sure the heading is not cut off from the body of the section.

To prevent heading-body separation problems, use one of these options:

- Select the heading. Click **Home | Paragraph** dialog box launcher | **Line and Page Breaks** tab. Under **Pagination**, check **Keep with next**.
- Insert a hard page break (Ctrl+Enter) to the left of the heading to force the heading onto the next page with the body.

Figures and tables must not be separated from their captions. If necessary, insert a hard page break (Ctrl+Enter) to the left of the table caption (which, by convention, is typed above the table) or the figure itself (the figure caption goes below the figure).

Errant blank pages. To delete blank pages in the middle of a document, go to the blank page, click it, and display the hidden symbols (**Home | Paragraph | Show/Hide ¶**). Delete spaces and paragraph symbols until the page is gone. Similarly, if the blank page is at the end of the document, press Ctrl+End and remove the hidden symbols to delete the extra page.

Document inspector

To check that you've removed all comments and tracked changes, click **File | Info | Check for Issues | Inspect Document**. Rather than selecting **Remove All**, click **Close**. Find the stray revision marks by clicking **Review | Changes | Next** and take action, if necessary, and delete or reject them.

Finally, print a hard copy

When you are confident that you've found and corrected all mistakes on-screen, print a hardcopy and proofread your paper again. Some mistakes are more easily identified on paper, and it is always better for you, rather than your instructor, to find them.

Good Housekeeping

Organizing your files in folders

You can expect to type at least one major paper and perhaps several minor writing assignments in every college course each semester. This amounts to a fair number of documents on your computer. One way to organize your files is to make individual course folders that contain subfolders for lecture notes, homework, and lab reports, for example. To create a new folder in Word when you are saving a file:

1. Click **File | Save As**. The **Save As** dialog box appears.
2. Click a link in the breadcrumb trail in the box at the top or use the navigation pane on the left to locate the folder in which you want to create a new folder.
3. Click the **New Folder** button. The **New Folder** dialog box appears.
4. Give the folder a short, descriptive name. Click **OK**.

To create a new folder in Windows Explorer:

1. Click **Start | Documents**. The **Documents library** dialog box appears.

2. Click **New Folder**. Type a name for the new folder and press **Enter**.

Accessing files and folders quickly

To make frequently used files or folders readily accessible, consider pinning them either to the **Recent** list in Word or to the **Favorites** list in Windows Explorer.

To pin files or folders to **Recent** in Word:

1. Click **File | Recent** to display recent documents and recent folders.
2. Right-click a file or folder and select **Pin to list** from the menu.
3. The file (folder) will be pinned to the top of the Recent Documents (Recent Places) list.
4. When you are finished working on the file (folder), click **File | Recent**, right-click the pinned file (folder), and select **Unpin from list**.

To pin files or folders to **Favorites** in Windows Explorer:

1. Open a Windows Explorer dialog box by clicking **Start | Documents**.
2. Browse your documents library until you find the frequently used file (folder).
3. Hold down the left mouse button and drag the file (folder) to **Favorites** in the navigation pane on the left (Figure A1.21).

Naming your files

File names can include letters, numbers, underlines, and spaces as well as certain punctuation marks such as periods, commas, and hyphens. The following characters are not allowed: \, /, :, *, ?, ", <, >, and |.

In general, file names should be short and descriptive so that you can easily find the file on your computer. If you intend to share your file, however, consider the person you are sharing it with. When you send your lab report draft to your instructor for feedback, put your name and the topic in the file name (for example, Miller_lacoperon); while "biolab1" may seem unambiguous to you, it is not exactly informative for your biology professor!

If you have forgotten the exact file name and location, you can search for it using Windows. Click **Start | Computer | Local Disk (C:)** (or **Documents** if you are able to narrow down the location) and search for files and folders.

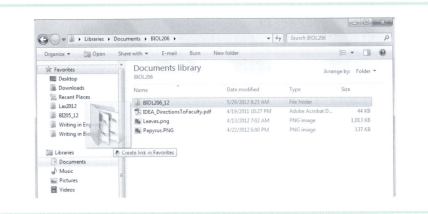

Figure A1.21 Pin frequently used files and folders to **Favorites** in Windows Explorer by clicking the file or folder and dragging it to the navigation pane on the left.

Saving your documents

When you write the first draft of a paper by hand, you have tangible evidence that you have done the work. When you type something on the computer, however, your work is unprotected until you save it. That means that if the power goes off or the computer crashes before you save the file, you have to start again from scratch. Hopefully it won't take the loss of a night's work to convince you to **save your work early and often**. Don't wait until you've typed a whole page; save your file after the first sentence. Then continue to save it often, especially when the content and format are complicated. Think about how long you would need to retype the text if it were lost and if you can afford to spend that much time redoing it.

Word automatically saves your file at certain intervals. You can adjust the settings by clicking **File | Options | Save | Save AutoRecover information every __ minutes**. It's also a good idea to save the file manually from time to time by clicking ∎ on the **Quick Access Toolbar**.

Backing up your files

It should be obvious that your backup files should not be saved on your computer's hard drive. Table A1.4 lists some offline and online backup options along with their advantages and disadvantages. In terms of offline options, USB flash drives are probably your best bet for compactness and convenience, but external hard drives come with software that lets you

TABLE A1.4   Possible options for backing up your electronic files

Backup Method	Capacity	Benefits	Drawbacks
External devices			
CD	700 MB	Portable	Not practical for large-scale backup
DVD	4.7 GB	No Internet required	
Blu-ray disc	Up to 128 GB	Not that expensive	Not reliable for long-term storage
USB 2.0 flash drive (jump drive, thumb drive)	Up to 128 GB		Few computers have blu-ray disc burners
MicroSD card	Up to 64 GB		Files cannot be accessed from mobile devices (*Exception*: Some mobile devices support SD cards and MicroSD cards)
SD card	Up to 128 GB		
External hard drive	Up to about 3 TB		
			If disc, drive, or card is lost, file is lost
Internal server	Depends on organization	Free	In-house Ethernet required
		Files can be shared or kept private	Internet and private network software required to access off-campus
			Limited storage capacity
Send emails to yourself	Depends on the email provider, but typically in the GB range	Free	Internet required
		Potentially infinite storage capacity	Limit to size of file that can be sent as attachment (typically 10–25 MB)
		Can be accessed from other computers and mobile devices	
Cloud storage	Infinite	All services offer a free storage option (2–7 GB)	Internet required
Box			Fees increase as storage needs increase
Dropbox		Can be accessed from other computers and mobile devices	Limit to file size that can be uploaded
Google Drive[a]			
Microsoft SkyDrive		Files can be shared or kept private	
SugarSync			

Source: Meece (2012) and Wynn (2012).

[a]Google Drive provides up to 5GB of free storage for non-Google formats (ppt, xls, jpg, mov, etc.) and unlimited for Google formats like Google Docs (D. Hiller, personal communication, 20 July 2010).

schedule automatic backups. With an account and an Internet connection, you can take advantage of any number of online options. These include Google Drive and other cloud services that store your files virtually, saving files to your organization's server, and even emailing files to yourself.

The most important thing about backing up your files is to have a plan. The odds are in your favor, unfortunately, that you will have a computer malfunction at least once every 3 years. For that reason alone

- Schedule regular, automatic backups for your hard drive (that way you won't forget)
- Have a second back up option for important files that you are currently working on (especially projects with tight deadlines)
- Make sure your backup options don't run out of storage space. As a rough guide, a 10-page text file has a size of about 200KB. In comparison, a picture taken with your digital camera is typically between 1000KB and 3000KB, and each of your music files is around 4000KB.

The bottom line is: Protect your valuable files with a reliable backup system.

Working with previous versions of MS Word

Starting with Word 2007, Microsoft introduced a new file format that decreases file size and increases security and reliability. The file extension of this new format is .docx; earlier versions of Word end with .doc. If you have an earlier version of Word, you will not be able to open .docx files unless you have installed the Compatibility Pack. To download this pack, type "office 2010 compatibility" into your browser's search box and download the Compatibility Pack from Microsoft's Home Page.

If Office 2007 or Office for Mac 2008 or later is installed on your computer, you will be able to open documents created in earlier versions of Word. After you open a .doc file, the title will appear with "Compatibility Mode" in parentheses on the title bar. To convert an older Word document to the new file format, click **File | Info | Convert**. A message warns you that the layout may change, but converting to the new file format allows you to take advantage of new features such as:

- Modifying imported Excel graphs directly in your Word document. In earlier versions, all corrections and modifications to graphs had to be made in Excel. In Word 2007 and later versions for both PCs and Macs, you *can* make changes to graphs after they have been imported into Word.

- Inserting equations by selecting a common equation from Word's list or typing your own equation into one of the built-in equation structures (see "The Insert tab, Equations").

- Checking for personal information or hidden content before sharing the document with others (see "The File tab"). This feature is handy for removing all comments and tracked changes on final versions of lab reports.

- Protected View to view files from potentially untrustworthy sources. You can manage the Trust Center settings by clicking **File | Options | Trust Center | Trust Center Settings**.

Finally, if you would like to share a Word 2007 document with someone who has an earlier version of Word and who has not installed the Compatibility Pack, click the **File** tab and select **Save As | Word 97–2003 Document**.

Making Graphs in Microsoft Excel 2010 and Excel for Mac 2011

Introduction

The commands for making graphs in Microsoft Excel 2010 for PCs and Excel for Mac 2011 are quite similar, but the user interface to access those commands is a little different. In this appendix, screenshots and instructions for both systems will be given unless the command sequence is the same, in which case Excel 2010 will be the default. Screenshots for Excel 2010 were taken on the Windows 7 operating system; those for Excel for Mac 2011 on the OS X operating system.

If you have been using Excel 2007, then Excel 2010 will not seem that different. However, if you are a Mac user, the Ribbon is a new addition to the user interface at the top of the screen. The **Ribbon** is a single strip that displays **commands** in task-oriented **groups** on a series of **tabs** (Figure A2.1). In Excel 2010, the new **File** tab replaces the Microsoft **Office Button** that was located in the top left of the screen in Excel 2007. Additional commands in some of the groups can be accessed with the **Dialog Box Launcher**, a diagonal arrow in the right corner of the group label. The **Quick Access Toolbar** comes with buttons for saving your file and undoing and redoing commands; you can also add buttons for tasks you perform frequently.

The window size and screen resolution affect what you see on the Ribbon. You may see fewer command buttons or an entire group abbreviated to one button on a smaller screen or if you have set your screen display to a lower resolution. Furthermore, in Excel 2010 you can create custom tabs on the Ribbon. The beauty of this function is that if you frequently use commands that are buried deep within menus and sub-menus, you can add

(A)   Quick access toolbar

Ribbon

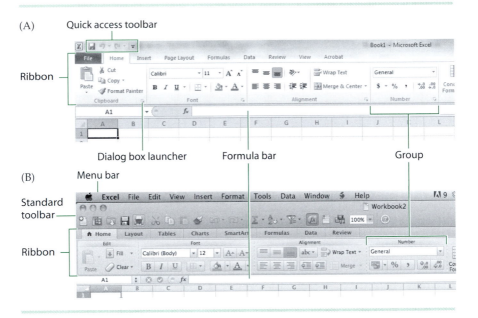

Dialog box launcher   Formula bar   Group

(B)   Menu bar

Standard toolbar

Ribbon

Figure A2.1   Screen display of part of the command area in (A) an Excel 2010 worksheet, and (B) an Excel for Mac 2011 worksheet.

these commands to your custom tabs for easy access. See the section "Commands in Excel 2010" for instructions on creating custom tabs. It is not possible to customize the Ribbon in Excel for Mac 2011, but the **menu bar** and the **standard toolbar** can be customized.

Handling computer files

The section "Good Housekeeping" in Appendix 1 applies equally to Word documents and Excel workbooks. Read over this section to develop good habits for naming, organizing, and backing up computer files. In addition, if you are making the transition from Microsoft Office 2003 (or earlier) to 2010, there are some things you should know about file compatibility, which is also covered under "Good Housekeeping."

Commands in Excel 2010

The best way to get up to speed with Excel 2010 is to learn where the commands you used most frequently in earlier versions of Excel are located. Table A2.1 is designed to help you with this task. Frequently used com-

TABLE A2.1	Common commands (listed alphabetically) and how to carry them out in Excel 2010	
Command	**Excel 2010 Command Sequence[a]**	**PC Hot Key**
Editing		
See Table A1.3		
File		
See Table A1.3		
Formatting		
Also see Table A1.3		
Cells	Home | Cells | Format | Format Cells	Ctrl+1 or Alt+HOE
Column width	Home | Cells | Format | Column Width	Alt+HOW
Numbers		
General, currency, date, etc.	Home | Number	Alt+HN or Alt+HFM
Increase or decrease decimal	Home | Number | Increase (Decrease) Decimal	Alt+H0 (increase) or Alt+H9 (decrease)
Paste Special	Home | Clipboard | Paste | Paste Special	Ctrl+Alt+V or Alt+HVS
Row height	Home | Cells | Format | Row Height	Alt+HOH
Sort data	Data | Sort & Filter | Sort	Alt+AS
	Home | Editing | Sort & Filter	Alt+HS
Sub- and superscript	Home | Font dialog box launcher | Effects | Sub- or Superscript	Alt+HFN
Symbols	Insert | Symbols | Symbol	Alt+NU
Page Layout		
Headers & Footers	Insert | Text | Header & Footer	Alt+NH
Insert cells, columns, rows, worksheets	Home | Cells | Insert	Alt+HI
Cells		Ctrl+Shift+=
Worksheet		Shift+F11

[a]Many of the Word 2010 commands can also be accessed by selecting text or an object and right-clicking. For Alt commands, hold down **Alt** while pressing the first letter key, which corresponds to a tab name File Home Insert Page Layout References Mailings Review View Add-Ins Acrobat . Then release **Alt** and type the next letter(s) in the sequence. F11 is a function key.

TABLE A2.1 *Continued*		
Command	**Excel 2010 Command Sequence**[a]	**PC Hot Key**
Insert charts (graphs)	Insert | Charts	Alt+N
Margins	Page Layout | Page Setup | Margins	Alt+PM
Orientation	Page Layout | Page Setup | Orientation	Alt+PO
Page break	Page Layout | Page Setup | Breaks	Alt+PB
Page numbers	Insert | Text | Header & Footer | Header & Footer Elements | Page Number	Alt+NH
Scaling	Page Layout | Page Setup | Size | More Paper Sizes	Alt+PSZM
Set print area	Page Layout | Page Setup | Print Area | Set Print Area	Alt+PRS
Print gridlines	Page Layout | Page Setup | Print Titles | ☑ Gridlines	Alt+PI
Repeat row or column titles on multiple-page worksheets	Page Layout | Page Setup | Print Titles	Alt+PI
View		
Headers and Footers	View | Workbook Views | Page Layout	Alt+WP
Page breaks	View | Workbook Views | Page Break Preview	Alt+WI
Print preview	File | Print	Alt+FP
Freeze (keep row(s) or column(s) fixed while scrolling through the rest of the worksheet)	View | Window | Freeze	Alt+WF

mands are listed in alphabetical order in the first column, and the Excel 2010 command sequence is given in the second column. The PC keyboard shortcuts for most of the commands are shown in the third column. If you are a Mac user,

Ctrl = Command (⌘)

Alt = Option (however, some Alt key sequences do not have a Mac equivalent)

In this book, the nomenclature for the command sequence is as follows:

Ribbon tab | Group | Command button | Additional Commands (if available).

For example, to make an XY graph, **Insert | Charts | Scatter | Scatter with only Markers** means "Select the **Insert** tab on the Ribbon, and in the **Charts** group, click the down arrow on the **Scatter** button, and select the **Scatter with only Markers** options."

For an overview of what's new in Excel 2010, go to Microsoft's Home Page (http://office.microsoft.com) and type "what's new in excel 2010" into the search box. The link for Excel for Mac 2011 is http://www.microsoft.com/mac/excel.

Formulas in Excel

Excel is a popular spreadsheet program in the business world, but its "number crunching" capabilities make it a powerful tool for data reduction and analysis in general. You can use Excel like a calculator by typing numbers and mathematical operators into a cell and then pressing **Enter**. Most likely, however, you will write your own formulas or choose common functions from Excel's collection and apply them to cell references instead of numbers. Excel is great for doing repetitive calculations quickly.

In this section you will learn how to write some formulas frequently used in biology. Even when you've become proficient at writing formulas in Excel, however, it's still a good idea to *do a sample calculation by hand (using your calculator) to make sure the formula you entered in Excel gives you the same result.* If you find a discrepancy, check the formula and check your math. Make sure the answer makes sense.

Formulas in Excel always start with an equal sign (=) followed by the cell references, numbers, and operators that make up the formula. Some commonly used operators are shown in Table A2.2. Excel performs the calculations in order from left to right according to the same order of operations used in algebra: first negation (−), then all percentages (%), then all exponentiations (^), then all multiplications and divisions (* or /), and finally all subtractions and additions (− or +). For example, the formula "= 100–50/10" would result in "95", because Excel performs division before subtraction. To change the order of operations, enclose part of the formula in parentheses. For example, "=(100–50)/10" would result in "5."

Writing formulas

Tasks you will encounter when writing formulas include entering a formula in the correct format or selecting a function from Excel's collection and applying the function to a range of cells.

TABLE A2.2	Operators commonly used for calculations in Excel
Operator	**Meaning**
+	Addition
-	Subtraction or negation
*	Multiplication
/	Division
%	Percentage
^	Exponentiation
:	Range of adjacent cells
,	Multiple, non-adjacent cells

TYPE A FORMULA IN THE ACTIVE CELL

1. Click a cell in which you want the result of the formula to be displayed (the so-called active cell). The selected cell will have a dark, black border, as in cell O4 in Figure A2.2A. The **Formulas** tab in Excel for Mac 2011 is shown in Figure A2.2B.
2. Type "=" (equal sign).
3. Type the constants, operators, cell references, and functions that you want to use in the calculation. See Table A2.3 for examples.
4. Press **Enter**. The result of the calculation is displayed in the active cell. You can view and edit the formula on the formula bar.

APPLY A FUNCTION TO VALUES IN ADJACENT CELLS

1. Click a cell in which you want the result of the formula to be displayed.
2. Type "=", the name of the function (e.g., AVERAGE) and then "(".
3. Hold down the left mouse button and select the cells you want to average. If there are a lot of cells to average, click the first cell, hold down the **Shift** key, and then click the last cell in the range. Regardless of the selection method, Excel automatically inserts the first cell and last cell of the range, separated by a colon, in the formula bar (see Figure A2.2A).
4. Type ")" and press **Enter**.

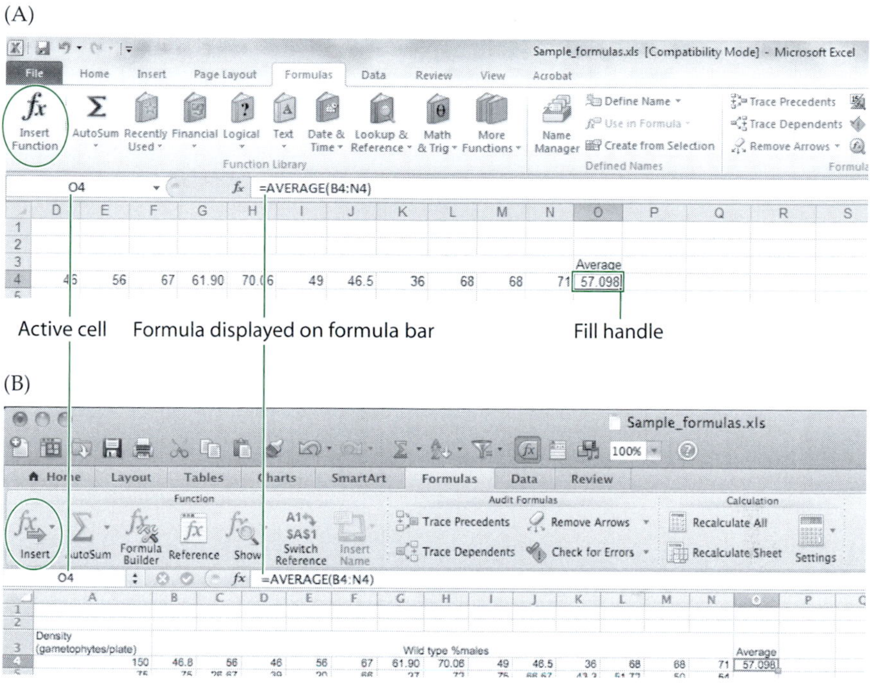

(A)

Active cell   Formula displayed on formula bar   Fill handle

(B)

Figure A2.2 The **Formulas** tab has an **Insert Function** button and other commands for doing calculations in (A) Excel 2010, and (B) Excel for Mac 2011.

APPLY A FUNCTION TO VALUES NOT IN ADJACENT CELLS

1. Click a cell in which you want the result of the formula to be displayed.

2. Type "=", the name of the function (e.g., AVERAGE) and then "(".

3. Hold down the **Ctrl** key and click the cells you want to average. Excel automatically inserts the selected cells, separated by commas, in the formula bar.

4. Type ")" and press **Enter**.

TABLE A2.3	Examples of formulas written in Excel
Formula	**Meaning**
=16*31-42	Assigns the product of 16 times 31 minus 42 to the active cell
=A1	Assigns the value in cell A1 to the active cell
=A1+B1	Assigns the sum of the values in cells A1 and B1 to the active cell
=(O4/H4)^(10/(I4-B4))	Value in cell O4 is divided by the value in cell H4. This number is raised to the power of 10 divided by the difference between the values in cells I4 and B4.
=((J5-K5)*60*L5)/(M5*N5)	Value in cell J5 minus the value in cell K5 is multiplied by 60 and multiplied by the value in cell L5. This result is divided by the product of the values in cells M5 and N5.
=SUM(A1:A26)	Assigns the sum of the values in cells A1 through A26 inclusive to the active cell
=SUM(B4,J4,L4,T4)	Assigns the sum of the values in cells B4, J4, L4, and T4 to the active cell
=AVERAGE(B4:U4)	Assigns the average of the values in cells B4 through U4 inclusive to the active cell
=AVERAGE(B4,J4,L4,T4)	Assigns the average of the values in cells B4, J4, L4, and T4 to the active cell
=log(B1)	Assigns the log of the value in cell B1 to the active cell
=ln(B1)	Assigns the natural log of the value in cell B1 to the active cell

SELECT AN EXCEL FUNCTION WITH THE INSERT FUNCTION BUTTON

1. Click a cell in which you want the result of the formula to be displayed (the active cell).
2. Click **Formulas | Function Library | Insert Function** (Figure A2.3A). Or, from any tab, click the **Insert Function** button to the left of the formula bar.
3. If the function you're looking for is not shown in the Recently Used category, click the down arrow to display another category.
4. When you locate the function you want to use, click it in the **Select a function** list box and then click **OK**.

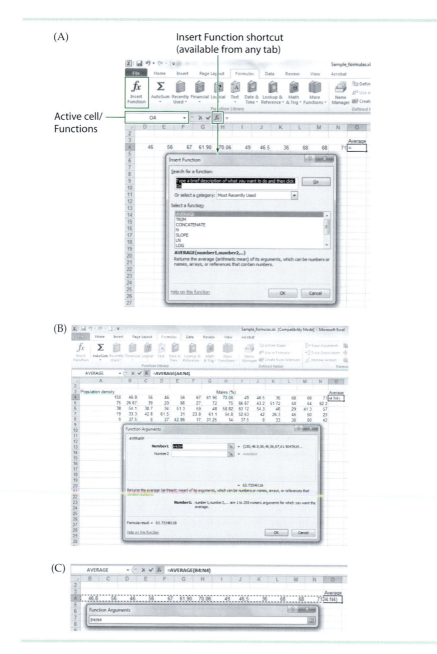

Figure A2.3   (A) Selecting a function from Excel's **Function Library**, and (B) applying it to a range of cells by typing the cell references separated by a colon. (C) Instead of typing the cell references (and potentially introducing typos), click the first cell, hold down the **Shift** key, click the last cell, and then hit **Enter**.

5. In the **Function Arguments** dialog box (Figure A2.3B), Excel tries to guess the range of cells to which you want to apply the function. If the range is incorrect, use the **Shift** key or the **Ctrl** key as explained previously to select the correct range (Figure A2.3C).

6. Click **OK**.

SELECT AN EXCEL FUNCTION BY TYPING AN EQUAL SIGN IN AN ACTIVE CELL

1. Click a cell in which you want the result of the formula to be displayed (the active cell).

2. Type "=" in the active cell. The most recently used function is displayed in the **Active Cell/Functions** box to the left of the formula bar (see Figure A2.3A).

3. Click the down arrow to select a function from the drop-down menu. The **Function Arguments** dialog box appears (see Figure A2.3B).

4. Follow steps 5 and 6 above.

Copying formulas using the fill handle

Quite frequently you may want to perform the same calculation on data contained in cells of neighboring rows or columns. Instead of copying and pasting the formula, you can simply drag the formula into adjacent cells. To do so, follow these steps:

1. Click the cell containing the formula you wish to copy.

2. Locate the fill handle, which is a small black square on the bottom right corner of the cell (see Figure A2.2A).

3. Move the mouse over the fill handle to display cross hairs (+).

4. Hold down the left mouse button and drag the fill handle over the cells you want to "fill" with the formula. For the sample data in Figure A2.3B, you would copy the formula into cells O5 through O8. These cells would then display the average of the cells in rows 5 through 8, respectively.

Copying cell values, but not the formula

When you select Copy, Excel copies everything in the cell—the formula, the number, and the text. When you select Paste, however, you may get a "#REF!" error instead of the entry you expected. This error typically occurs when the connection between the cell references and the formula is lost. To paste the value without the formula, click **Home | Clipboard | Paste | Paste Values**.

Unfurling the Ribbon

If you prefer a more visual approach to the commands, the following sections show you the tabs you're likely to use for entering data, plotting graphs, formatting a spreadsheet before you print it, and selecting the way your document appears on the screen. In addition to the buttons on the tabs, some groups have a diagonal arrow that launches a dialog box for those particular tasks (Excel 2010 but not Excel for Mac 2011). The dialog boxes are similar (if not identical) to those in Excel 2003 and 2007.

The File tab

Clicking the **File** tab opens a new dialog box called the **Backstage View** (Figure A2.4). The left panel lists the file-related commands, the center panel provides the options for that command, and the right panel shows a preview or other options. Document commands such as Save, Save As, Open,

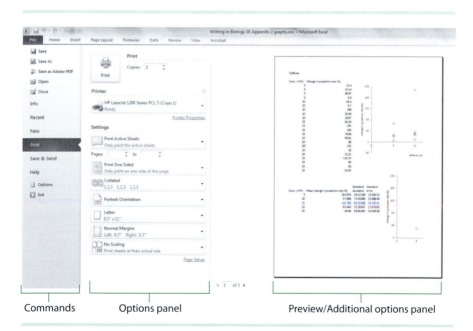

Commands   Options panel   Preview/Additional options panel

Figure A2.4   This Backstage View of the **File** tab consists of three panels: commands, options, and preview. Workbook-related commands such as Save, Save As, Open, Close, and Print are shown on the left panel. The options panel shows additional commands related to the command selected on the left panel, in this case **Print**. The preview panel displays the page as it will be printed. For commands other than **Print**, additional options are shown on the right panel.

Close, Recent, New, and Print, which were located on the **Office Button** in Word 2007 and on the **File** menu in Word 2003, are located on the **File** tab. Other commands on the **File** tab are

- **Info**, which provides new file management tools for setting permissions, inspecting the document for hidden properties and personal information, and recovering earlier autosaved or unsaved versions of a file.

- **Save & Send** makes it easier to collaborate with others on files. All collaborators must use Office 2010 and possess SharePoint 2010 software or a SkyDrive account to access and save the files online. Similar to Google docs, this feature allows two or more people to work on the same document at the same time.

- **Help**. The **Help** menu can also be accessed by pressing the **F1** function key.

- **Options** allows you to change program settings such as user name, how formulas are calculated and how errors are checked, AutoCorrect options, how documents are saved and printed, and so on. The **Options** button is also the gateway to commands for customizing the Ribbon and the Quick Access Toolbar and for managing Microsoft Office add-ins and security settings.

Customize the Ribbon. This feature lets you create a new tab on the Ribbon, to which you can add hard-to-find commands that you use frequently. Click **File | Options | Customize Ribbon**. After clicking the **New Tab** button, add the desired groups and commands to the new tab (see Figure A1.6).

The Home tab

The **Home** tab contains word-processing functions such as copy, cut, and paste; commands for formatting text; and the familiar **Find** and **Replace** buttons (Figure A2.5). In addition, there are commands for formatting cells and sorting data. From left to right in Figure A2.5, these command buttons are **Wrap Text**, **Increase/Decrease Decimal**, **Format Cells**, and **Sort & Filter**.

Wrap text. When you type a long text in a cell, it runs into the adjacent cell to the right. If there is text or a numerical entry in the adjacent cell, the long text is hidden (it appears to be cut off). To make the long text visible, you could widen the columns, but then fewer columns will be visible on the screen. To be able to maintain column width and still view the entire text, click the desired cell and then **Home | Alignment | Wrap Text**. Excel expands the row height to accommodate the contents of the cell.

(A)

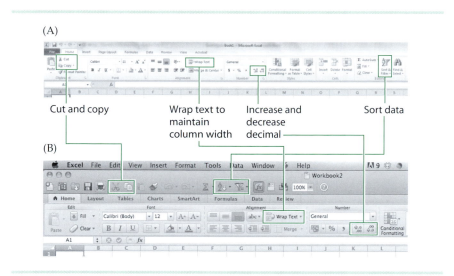

Cut and copy   Wrap text to   Increase and   Sort data
                      maintain   decrease
                      column width   decimal

(B)

Figure A2.5   Useful commands found on the **Home** tab of (A) an Excel 2010 worksheet, and (B) an Excel for Mac 2011 worksheet. The **Sort** and **Filter** commands are also on the **Data** tab in Excel for Mac 2011.

Increase or decrease decimal.   When a calculation results in a value that has more decimal places than the measurements from which it originated, round off. In other words, display the same number of significant figures as the smallest number of significant figures in any measurement. To round off, click the desired cell and then **Home | Number | Decrease Decimal**.

Format cells. Home | Cells | Format Cells opens a dialog box with 6 tabs: Number, Alignment, Font, Border, Fill, and Protection. The options selected can be applied to one or many cells.

Sort data.   If you'd like to alphabetize a list of names or arrange numerical values in ascending or descending order, use the **Home | Editing | Sort & Filter** command. First, select *all of the columns that contain data* that you want to sort. This is important, because if you select only one column, then only the data in that column will be sorted. Data in the adjacent columns will then no longer correspond to the correct data in the sorted column. After selecting all of the data, click **Custom Sort** on the **Sort & Filter** button's drop-down list. Select the criteria according to which you want to sort: by column, by values, and in which order. Then click **OK**.

The Insert tab

In Excel 2010, the **Insert** tab has commands for inserting tables, illustrations, charts, links, text, and symbols (Figure A2.6A). **Insert | Charts** is where you begin the process of making graphs for your lab reports. In Excel for Mac 2011, the commands for inserting tables and charts are on separate tabs.

(A)

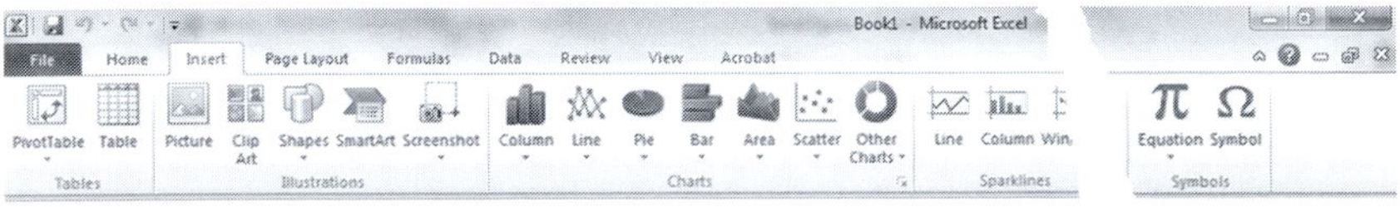

Excel 2010

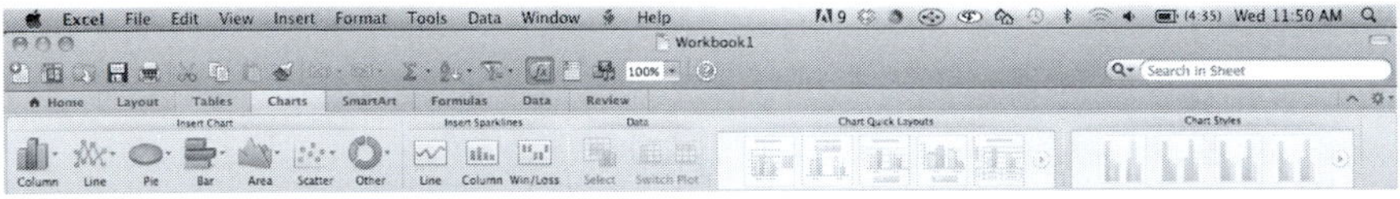

Excel for Mac 2011

(B)

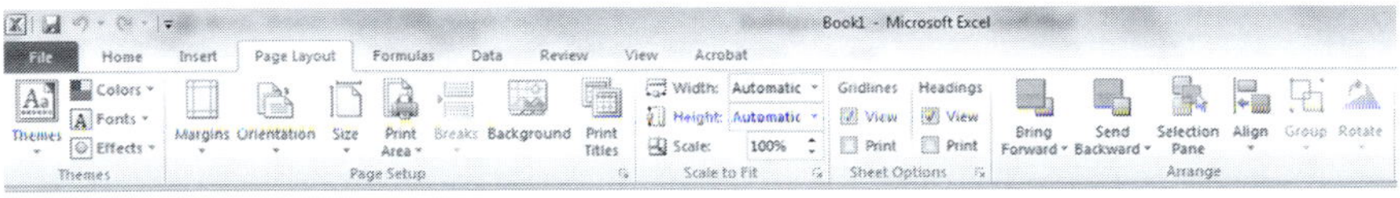

Excel 2010

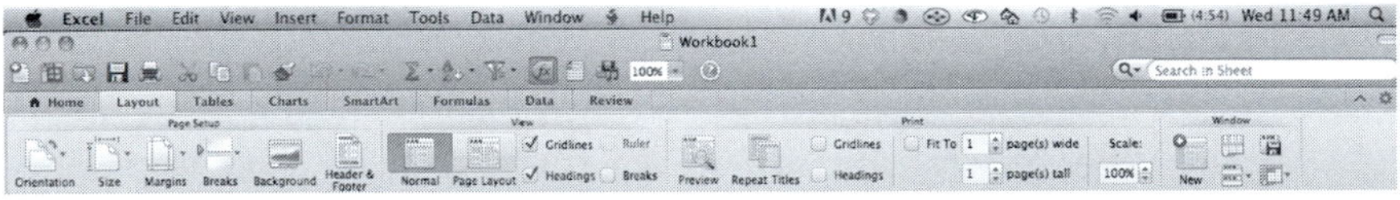

Excel for Mac 2011

(C)

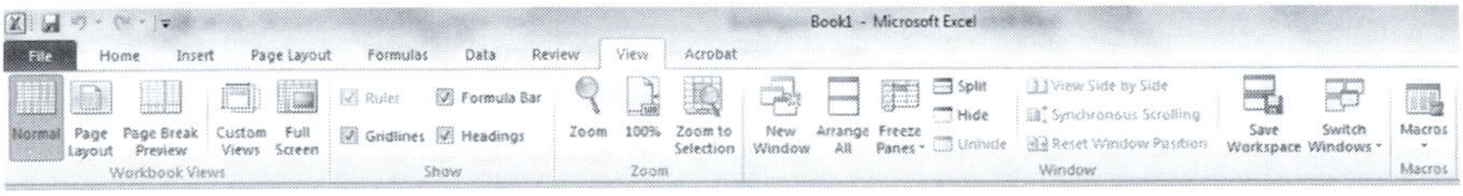

Excel 2010

Figure A2.6 (A) The **Insert** tab (Excel 2010) and the **Charts** tab (Excel for Mac 2011) have commands for graphing data. (B) **Page Layout** buttons are used for formatting the spreadsheet. (C) The **View** tab affects how you see your worksheet on the screen. Some of the buttons on the Excel 2010 **View** tab are found on the **Layout** tab in Excel for Mac 2011.

Detailed instructions for making XY graphs, vertical and horizontal bar graphs, and pie charts are given in the following sections.

The Page Layout tab

The **Page Layout** tab is where you format your spreadsheet after you've entered data (Figure A2.6B). The buttons on this tab allow you to

- Change the margins
- Add headers, footers, or page numbers
- Scale the content to fit on one sheet of paper
- Adjust page breaks
- Set the print area
- Repeat row or column titles on multiple-page worksheets

The View tab

The **View** tab in Excel 2010 gives you options for displaying your workbook(s) on the screen (Figure A2.6C). In Excel for Mac 2011, some of these buttons are found on the **Layout** tab. There are three workbook views that can also be accessed from the status bar (Figure A2.7).

- **Normal** view is the default view and is typically used when entering data.

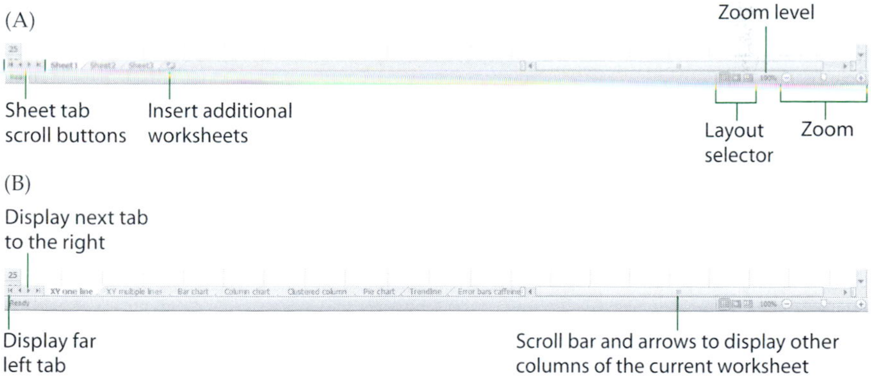

(A)

Zoom level

Sheet tab
scroll buttons

Insert additional
worksheets

Layout
selector

Zoom

(B)

Display next tab
to the right

Display far
left tab

Scroll bar and arrows to display other
columns of the current worksheet

Figure A2.7 Customizing, inserting, and navigating between and within worksheets. (A) Excel 2010 default workbook with three worksheets. (B) Excel 2010 workbook with extra, customized worksheet tabs.

- **Page Layout** view displays margins; headers, footers, and page numbers; and page breaks.
- **Page Break Preview** is handy for adjusting page breaks to keep blocks of text or data together

When a spreadsheet contains a lot of data, the row and column titles disappear when you scroll down or to the right. To be able to view the titles as you scroll through the worksheet, click **View | Window | Freeze Panes | Freeze Top Row** (or **Freeze First Column**). **Split** view splits the worksheet into 2 or 4 sections (based on the cell that is selected), each of which can be scrolled independently. Use **Split** view to view different sections of the worksheet simultaneously. Use **Freeze Panes** specifically to lock row and column titles while scrolling through the rest of the worksheet.

More tabs below

Each Excel 2010 workbook comes with three worksheets, named Sheet1, Sheet2, and Sheet3 (Figure A2.7A). Each sheet has its own tab, located in the bottom left corner of the screen, just above the status bar. You can customize the tabs for each sheet by double-clicking the name on the tab and typing a better descriptor (Figure A2.7B). You can also rearrange the worksheets by dragging a tab to a different position. Finally, if you'd like to add more worksheets, click the **Insert Worksheet** button, located to the right of the last sheet tab. Some good reasons for keeping multiple worksheets in one workbook are:

- To collect replicate data, yet keep individual trials separate
- To collect data from multiple lab sections for the same experiment, to pool or keep separate as needed
- To simplify file organization

To the left of the sheet tabs, you'll see four sheet tab scroll buttons. The scroll buttons are only needed if the workbook contains so many worksheets that their tabs cannot all be displayed at once, as in Figure A2.7B. In that case, the first scroll button causes the far left worksheet tab to be displayed, the second button displays the next tab to the left, the third button displays the next tab to the right, and the fourth button displays the far-right worksheet tab.

The bottom right corner of the workbook has three features related to the view: Layout Selector, Zoom Slider, and Scroll bar and arrows. The **Layout Selector** is similar to the one in Excel 2003, in which you can select among three layouts: Normal, Page Layout View, and Page Break Preview. The **Zoom Slider** lets you zoom in or out on certain cells in the worksheet. The horizontal scroll bar and arrows let you scroll across columns; the vertical scroll bar and arrows allow you to move across rows. Alternatively, if your mouse has a wheel, rotate the wheel to navigate vertically or hold down the wheel and move the mouse horizontally until the desired row or column is displayed.

Excel Terminology

Every discipline has its own terminology and the Microsoft-dominated computer world is no exception (Table A2.4). In the sections that follow, I use primarily Excel terminology to make it easier for you to find the buttons for making and formatting graphs. Keep in mind, however, that this is not really the language of scientists!

Goodbye Chart Wizard, Hello Insert Charts

In Excel 97 through 2003, Chart Wizard prompts you to select a chart type, confirm the arrangement of the source data, input chart and axis titles, and select where in the workbook the graph should be inserted. In Excel 2007 and 2010, Chart Wizard has been replaced by chart-specific command buttons on the **Insert** tab (see Figure A2.6A).

TABLE A2.4	Excel-specific terminology
Excel Term	**Description**
Workbook	An Excel file that initially consists of three worksheets
Worksheet	Spreadsheet
Chart	Graph
Charting	Plotting
Scatter chart	Line graph or XY graph
Line chart	Not a line graph; do not use this type of chart
Column chart	Vertical bar graph
Bar chart	Horizontal bar graph
Data series	Set of related data points
Plot area	Area of the graph inside the axes
Chart area	Area outside the axes but inside the frame
Legend	Legend or key
Marker	Data symbol

Plotting XY Graphs (Scatter Charts)

Scientists call XY graphs "line graphs," but you should not confuse line graphs with Excel's "line charts." Line charts do not space data proportionally on the x-axis. For example, intervals of 5, 20, and 50 units would be spaced equally, when in fact there should be 5 units in the first interval, 15 in the second, and 30 in the third. The bottom line is: *When you want to make a line graph in Excel, choose "scatter charts."*

Entering data in worksheet

Before you enter data in an Excel worksheet, you must first have a clear idea of what your XY graph should look like. Which parameter should be plotted on the x-axis and which one on the y-axis? By convention, the x-axis of the graph shows the independent variable, the one that was manipulated during the experiment. The y-axis of the graph shows the dependent variable, the variable that changes in response to changes in the independent variable.

On the worksheet, Column A is used to enter the data for the x-axis, whereas subsequent columns are used for data for the y-axis.

Let's say you carried out an experiment in which you tested the activity of an enzyme (catalase) at different temperatures (Figure A2.8). Because temperature is the variable you manipulated, enter the temperature data in Column A. Enter the catalase activity data in Column B. For a simple graph like this with only one dataset, it is not necessary to provide column headings, although for future reference it is always good to include the variable name and the units for the data.

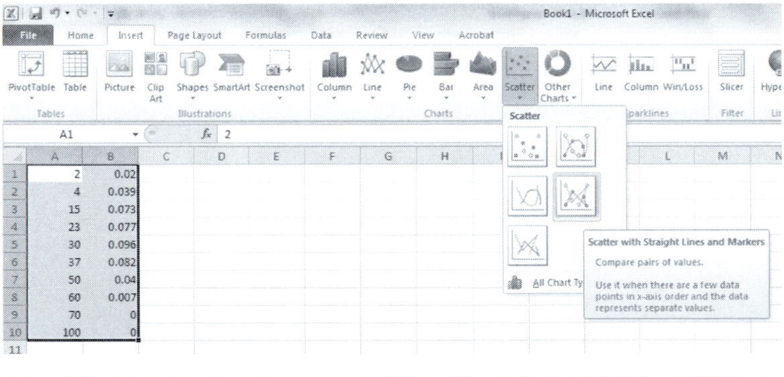

Figure A2.8 Select **Scatter**, not **Line**, to make an XY graph. The chart type is selected with **Insert | Charts** in Excel 2010; **Charts** in Excel for Mac 2011. Enter the data for the independent variable (x-axis) in Column A. Enter the data for the dependent variable (y-axis) in Column B.

Creating the scatter chart

1. Hold down the left mouse button and select the cells containing the values you want to plot. If there are a lot of cells, click the first cell, hold down the **Shift** key, and then click the last cell in the range. Note: If the data to be plotted are not in adjacent columns, select the first column, hold down the **Ctrl** key, and then select the data in any other column(s).

2. In Excel 2010, click **Insert | Charts | Scatter | Scatter with Straight Lines and Markers** (see Figure A2.8). Holding the mouse over each type of scatter chart opens a pop-up window with a description of that type of graph. Never use Line Chart, as this option spaces the x-axis values at equal intervals, instead of according to the intervals of the data. In Excel for Mac 2011, click **Charts | Scatter | Straight Lined Scatter**.

3. Clicking the plot area (the area inside the axes) or the chart area (the area outside the axes) of the newly created graph activates the **Chart Tools** contextual tab, which has three tabs of its own (Figure A2.9). Click the **Chart Tools Layout** tab to view the but-

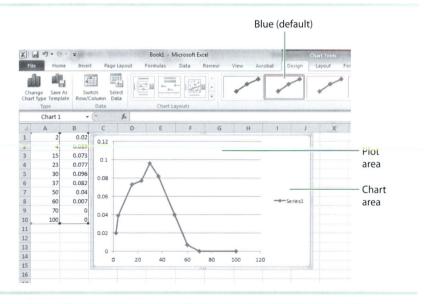

Figure A2.9 The **Chart Tools** tab is displayed in Excel 2010 when clicking the plot area or chart area of the graph. One of the three tabs associated with the **Chart Tools** tab is the **Chart Tools Design** tab, which has commands for changing the chart type, saving the chart format as a template, and changing the color of the points and lines.

(A)

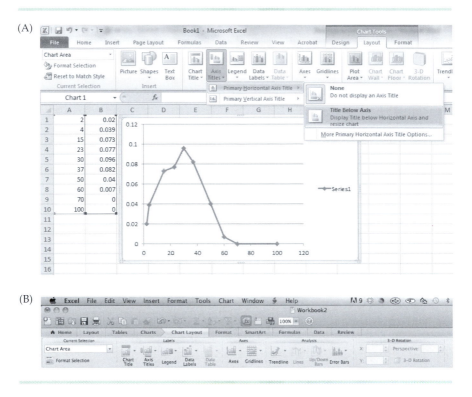

(B)

Figure A2.10 (A) To add a text box for the *x*-axis title in Excel 2010, click **Axis Titles | Primary Horizontal Axis Title | Title Below Axis** on the **Chart Tools Layout** tab. (B) In Excel for Mac 2011, the same commands are found under **Chart Layout | Axis Titles**.

tons for formatting the graph (Figure A2.10A). In Excel for Mac 2011, the same commands are found under **Chart Layout** (Figure A2.10B). Do not insert a Chart Title unless you are preparing this graph for an oral presentation.

4. In Excel 2010, to add a text box for the *x*-axis title, click **Axis Titles | Primary Horizontal Axis Title | Title Below Axis** (see Figure A2.10A). In Excel for Mac 2011, the same commands are found under **Chart Layout | Axis Titles** (see Figure A2.10B).

In the present example, the *x*-axis title is "Temperature (°C)." In Excel 2010, type "Temperature (". Then switch to the **Insert** tab and click **Symbols | Symbol** (Figure A2.11A). Click the degree

(A)

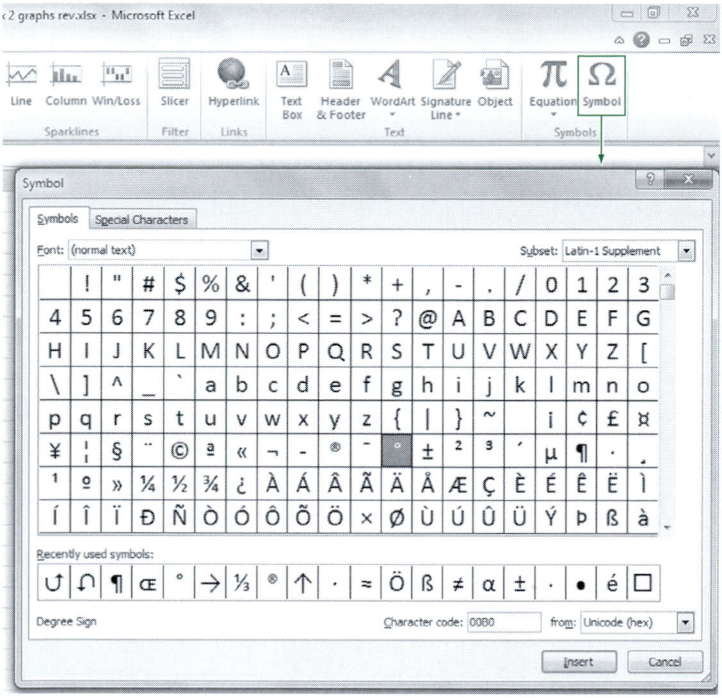

(B)

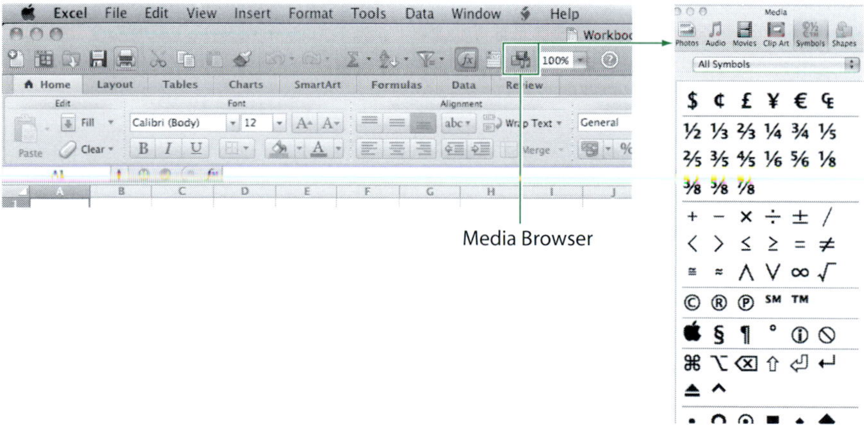

Media Browser

Figure A2.11   (A) In Excel 2010, insert a symbol in the axis title by clicking **Insert | Symbols | Symbol**. (B) In Excel for Mac 2011, click the **Media Browser** icon on the standard toolbar and then **Symbols** or click **Insert | Symbol...** on the menu bar.

symbol, click **Insert**, and then click **Close**. Then type "C)".
Unfortunately, it is not possible to make shortcut keys or to program symbols in AutoCorrect in Excel 2010. In Excel for Mac 2011, click the **Media Browser** icon on the standard toolbar and then **Symbols** (Figure A2.11B) or **Insert | Symbol** on the menu bar.

5. In Excel 2010, to add a text box for the y-axis title, click **Axis Titles | Primary Vertical Axis Title | Rotated Title** (Figure A2.12). In Excel for Mac 2011, the same commands are found under **Chart Layout | Axis Titles** (see Figure A2.10B).

The y-axis title in the present example is "Catalase activity (product formed · sec^{-1}). This title contains both a symbol (middle dot) and a superscript ($^{-1}$). In Excel 2010, type "Catalase activity (product formed", space, click **Insert | Symbols | Symbol**, find the middle dot and select it, click **Insert**, and then click **Close**. In Excel for Mac 2011, select the middle dot using the **Media Browser** icon on the standard toolbar and then **Symbols** (see Figure A2.11B) or **Insert | Symbol** on the menu bar.

Continue typing the axis title by entering a space and then "sec–1)". In Excel 2010, to superscript the "–1," make sure you are typing in the *text box* for the y-axis title, *not* the formula bar. With the blinking cursor just ahead of the minus sign, hold

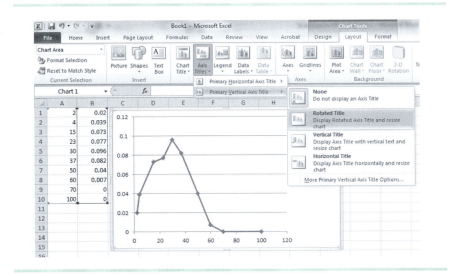

Figure A2.12 To add a text box for the y-axis title in Excel 2010, click **Axis Titles | Primary Vertical Axis Title | Rotated Title** on the Chart Tools Layout tab. In Excel for Mac 2011, the same commands are found under **Chart Layout | Axis Titles**.

down the **Shift** key and hit the up arrow key (↑) twice to select the "−1." Click **Home | Font** dialog box launcher and click the **Superscript** checkbox and then **OK** (Figure A2.13A).

(A)

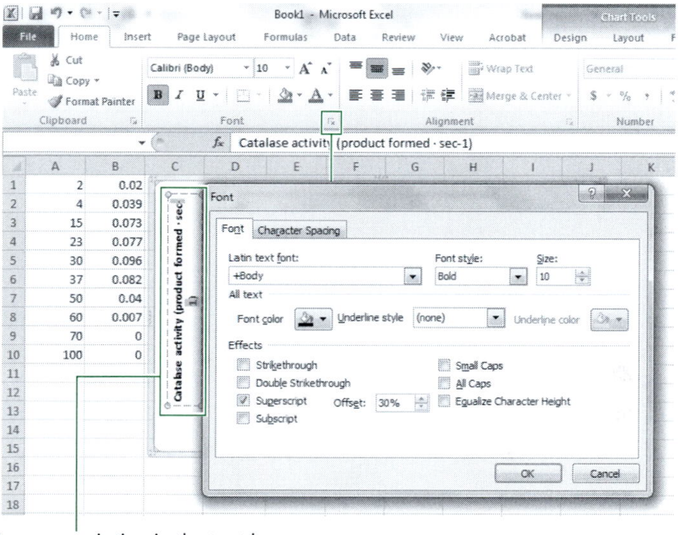

Select text for superscripting in the text box

(B)

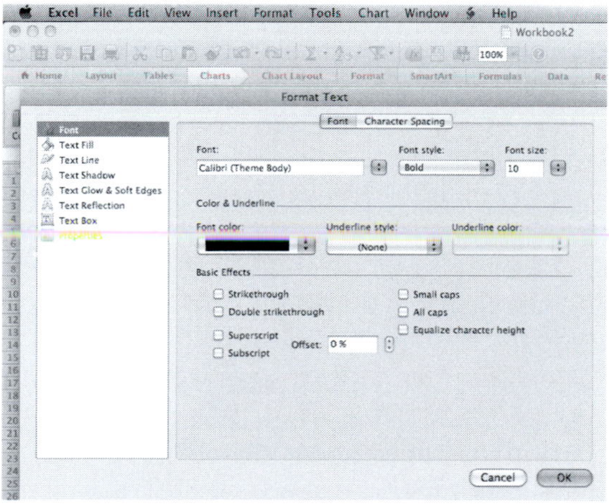

Figure A2.13   (A) In Excel 2010, superscript part of an axis title by selecting the text and clicking the **Home | Font** dialog box launcher and then **Superscript**. This command is only available when the blinking cursor is in the axis title text box, not in the formula bar. (B) In Excel for Mac 2011, select the text to be superscripted, right-click and select **Font** from the menu. In the **Format Text** dialog box, select **Superscript**.

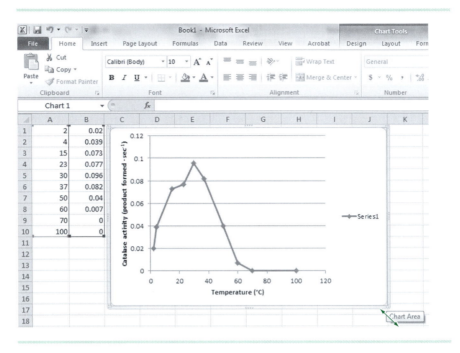

Figure A2.14   Enlarge the chart area to prevent a long *y*-axis title from wrapping.

In Excel for Mac 2011, select the "−1," right-click, and select **Font** from the menu. In the **Format Text** dialog box, select **Superscript** (Figure A2.13B).

If the axis title is longer than the axis itself, part of the title will wrap onto the next line. To avoid wrapping, simply enlarge the chart: Position the mouse pointer on the lower right corner of the graph so that it changes to a double-headed arrow (Figure A2.14). Drag the corner to enlarge the graph.

6. Because there is only one line on this graph, no legend is required. On the **Chart Tools Layout** tab, click the **Legend** button and select **None, Turn off Legend** (Figure A2.15). Alternatively, single-click the text box containing "Series 1" and hit the **Delete** key.

7. The data should fill the plot area so that there is no "dead space." The easiest way to eliminate the dead space beyond 100 on the *x*-axis is to double-click any number on the *x*-axis to open the **Format Axis** dialog box. Other ways to access this dialog

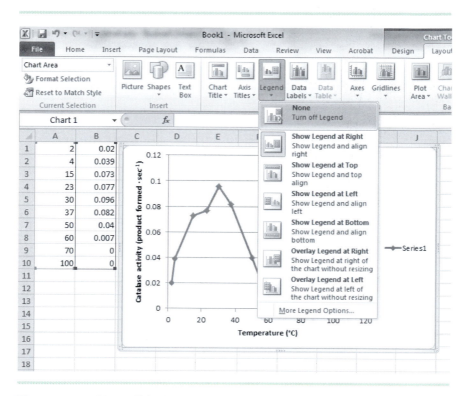

Figure A2.15 Turn off the legend when there is only one line on the graph. In Excel for Mac 2011, the same commands are found under **Chart Layout | Legend**.

box in Excel 2010 are to click **Axes | Primary Horizontal Axis | More Primary Horizontal Axis Options** (Figure A2.16A) or to right-click any number on the *x*-axis (Figure A2.16B). In the **Format Axis** dialog box, change the Maximum from **Auto** to **Fixed** and type "100". Then click **Close**.

In Excel for Mac 2011, double-click the *x*-axis or right-click and select **Format Axis | Scale** (Figure A2.16C). Simply enter the desired values for minimum, maximum, and major units (major units are the intervals).

8. Adjust the *y*-axis scale in a similar manner. Still in the same **Format Axis** dialog box, click **Number | Category: Number** and **Decimal places: 2** so that all of the numbers on the *y*-axis will display 2 decimal places.

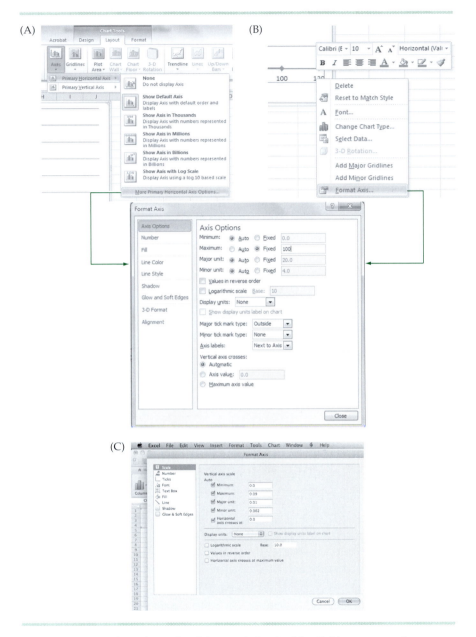

Figure A2.16 Adjust the scale of the *x*-axis by double-clicking any number on the *x*-axis to open the **Format Axis** dialog box. This box can also be accessed in Excel 2010 by (A) clicking the **Chart Tools Layout | Axes** button or (B) right-clicking a number on the *x*-axis. Change the **Maximum** by clicking **Fixed** and replacing 120 with 100. (C) In Excel for Mac 2011, open the **Format Axis** dialog box by double-clicking any number on the *x*-axis. Then enter the desired values for minimum, maximum, and major units (major units are the intervals).

9. The CSE Manual (2006) recommends using multiples of 1, 2, or 5 for the intervals on the axes. If the intervals have changed as a result of adjusting the minimum or maximum axis values, be sure to adjust **Format Axis | Axis Options | Major unit** accordingly.

10. The horizontal gridlines detract from the trend displayed by the data points. To remove the gridlines in Excel 2010, double-click one to open the **Format Major Gridlines** dialog box. Select **Line Color | No Line** and then **Close**. Alternatively, click **Chart Tools Layout | Gridlines | Primary Horizontal Gridlines | None, Do not display Horizontal Gridlines** (Figure A2.17). The command sequence is the same in Excel for Mac 2011.

11. The type of marker chosen by Excel—the diamond—is not among those recommended in the CSE Manual (2006) for research articles. The symbol hierarchies of Excel and the CSE Manual are compared in Table A2.5. The symbols that science editors use are based on ease of recognition and good contrast in black-and-white journal publications. Keep in mind that colored markers may look good on your computer screen, but they

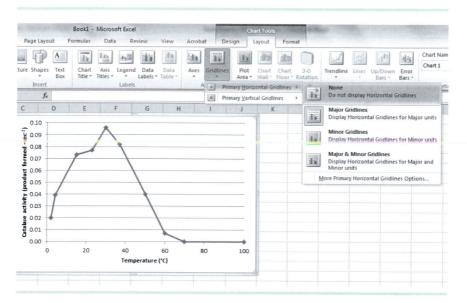

Figure A2.17   Turn off horizontal gridlines by double-clicking a gridline or using the **Chart Tools Layout | Gridlines** button.

TABLE A2.5	Comparison of Excel and CSE Manual symbol hierarchy for line graphs
Excel	**CSE Manual**
Blue diamond	Black open circle
Red square	Black filled circle
Green triangle	Black open triangle
Purple x	Black filled triangle
Turquoise x with additional vertical line	Black open square
Brown circle	Black filled square

will appear as various shades of gray when you print out your graph on a black-and-white printer.

a. To change the diamond to an open circle, double-click one of the markers or right-click a marker and select **Format Data Series** (Figure A2.18A).

b. Under **Marker Options | Marker Type**, click the **Built-in** option button. This selection allows you to choose the circle from the drop-down menu and to adjust the size of the marker.

c. Now click **Marker Fill | No fill** to make an open circle.

d. Under **Line Color**, click **Solid line** and choose black from the **Color** drop-down menu.

e. Under **Marker Line Color**, click **Solid line** and choose black from the **Color** drop-down menu.

f. Finally, click **Close**.

In Excel for Mac 2011, double-click or right-click one of the markers to open the **Format Data Series** dialog box, which contains the same options for formatting the markers as Excel 2010 (Figure A2.18B).

12. To remove the border around the graph, double-click anywhere in the chart area or right-click and select **Format Chart Area** from the menu. In Excel 2010, inside the **Format Chart Area** dialog box, click **Border Color | No line** and then **Close**. In Excel for Mac 2011, inside the dialog box click **Line | No Line**.

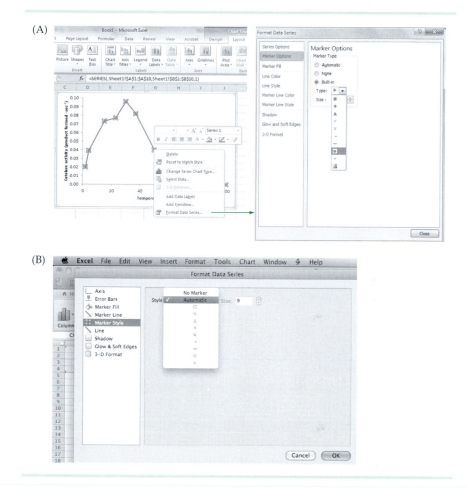

Figure A2.18   Change the marker type and fill by double-clicking a marker or by right-clicking a marker and selecting **Format Data Series**. (A) In Excel 2010, under **Marker Options**, click **Built-in** and choose a circle, triangle, or square from the drop-down menu. (B) In Excel for Mac 2011, these options are found under **Format Data Series | Marker Style.**

The final graph, formatted according to CSE Manual recommendations, looks like Figure A2.19. This graph was copied and pasted from Excel 2010 into a Word 2010 document. Further changes can be made to the graph in Word 2010. Simply double-click on the element and make the appropriate revisions in the dialog box.

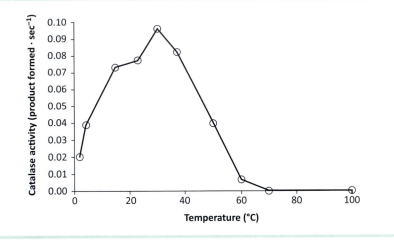

Figure A2.19   Final form of scatter chart after formatting in Excel. Additional changes can be made in a Word document (2007 or later) after copying and pasting the graph from Excel.

Save chart formatting

XY graphs are very common in biology and they all have basically the same format. To apply the format described above to XY graphs that you make in the future, see the section on "Making Chart Templates" on pp. 257–260.

Adding Data after Graph Has Been Formatted

If you have gone to all this trouble to format a graph, and realize after the fact that you need to add another data point, the last thing you want to do is to start from scratch. Fortunately, Excel 2010 makes it possible to add data after the graph has been made.

To incorporate additional data points in the same series

Insert a new row in the worksheet for the new data points. To do this, click a cell in the row below where the new row is to be inserted. Then click **Home | Cells | Insert | Insert Sheet Rows**. Enter the value for the x-axis in Column A and that for the y-axis in Column B. The graph is automatically updated for these values.

To incorporate additional lines on the same graph

These instructions assume that the x-axis values remain the same and you would like to insert additional y-axis values that represent data for a different treatment or condition. The y-axis values for the existing line are already in Column B. Type the y-axis values for the second line in Column C. For each additional line, enter data in the next column.

Now right-click anywhere in the chart area or the plot area of the existing graph and click **Select Data**.

1. In the **Select Data Source** dialog box, click the **Add** button in the **Legend Entries (Series)** list box.

2. In the **Edit Series** dialog box, under **Series name** click the cell reference containing the column heading or enter a short title for the legend. A legend is needed to distinguish the data sets when there are two or more lines on the graph.

3. Under **Series X values**, click the icon with the red arrow pointing to a cell in a worksheet. Select the values in Column A. Click the icon again. This action enters the range of values for the x-axis.

4. Under **Series Y values**, click the icon with the red arrow pointing to a cell in a worksheet. Select the values in Column C. Click the icon again. This action enters the range of the new values for the y-axis.

5. Repeat Steps 2–4 for each data set. Then Click **OK** to close the **Select Data Source** dialog box.

To change the legend titles

In earlier versions of Excel, the legend titles could be changed by simply double-clicking the text box containing the title. In Excel 2010, this method no longer works. Here are the new instructions:

1. Right-click a marker of the data series whose title you want to change. Click **Select Data** from the drop-down menu.

2. Click **Legend Entries (Series) | Edit**.

3. In the **Edit Series** dialog box, enter the new name for the series. Click **OK** twice.

Multiple Lines on an XY Graph

Plotting multiple lines on one graph is often the most efficient way to compare the results from several different treatments. How many lines should you put on one set of axes? The CSE Manual recommends no more than eight, but use common sense. You should be able to follow each line individually, and the graph should not look cluttered.

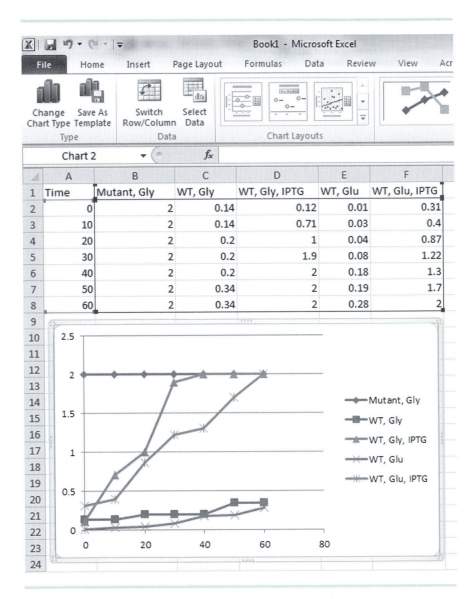

	A	B	C	D	E	F
1	Time	Mutant, Gly	WT, Gly	WT, Gly, IPTG	WT, Glu	WT, Glu, IPTG
2	0	2	0.14	0.12	0.01	0.31
3	10	2	0.14	0.71	0.03	0.4
4	20	2	0.2	1	0.04	0.87
5	30	2	0.2	1.9	0.08	1.22
6	40	2	0.2	2	0.18	1.3
7	50	2	0.34	2	0.19	1.7
8	60	2	0.34	2	0.28	2

Figure A2.20   Column A contains the *x*-axis data and Columns B through F contain the *y*-axis data representing different treatment conditions. Excel makes the legend on the graph from the column headings.

Entering data in worksheet

Type the data for the x-axis in Column A and the data for the dependent variable for each line in Column B, C, and so on. In the first row of each column, enter a short title that will be used in the legend to identify the different treatments or conditions.

Let's say you monitored the production of an enzyme called beta-galactosidase by *E. coli* grown under different conditions. Every 10 minutes for one hour, you measured the absorbance (which represents the concentration of beta-galactosidase) in each culture condition. Time is the independent variable that will be plotted on the x-axis. According to Excel convention, therefore, time is entered in Column A. The next five columns will contain the absorbance data at each time for the five different conditions (Figure A2.20).

Creating the scatter chart

1. Select the data you just entered in Columns A through F (including the headings) and click **Insert | Charts | Scatter | Scatter with Straight Lines and Markers** (Excel 2010) or **Charts | Scatter | Straight Lined Scatter** (Excel for Mac 2011).

2. Follow the instructions for plotting XY graphs, except Step 6, as described on pp. 220–232.

3. A legend *is* necessary to differentiate the five lines. However, CSE guidelines recommend that the legend be positioned inside the plot area when there is room. To move the legend inside the plot area, click the legend box so that selection handles appear. Move the mouse pointer over the legend to display crossed double-headed arrows and drag the legend to an open space inside the plot area.

The final graph, formatted according to CSE Manual recommendations, looks like Figure A2.21. Further changes to the graph can be made in Word 2010 after copying it from Excel 2010. Because XY graphs with multiple lines are very common in biology, consider saving the format in a chart template (see pp. 257–260).

Trendlines

Biologists often collect data to try to understand the relationship between two or more quantitative variables. If one of the variables explains or influences the other, then this so-called explanatory or independent variable is plotted on the x-axis. The variable that shows the response is plotted on the y-axis. In some observational studies, there may not be a causative relation-

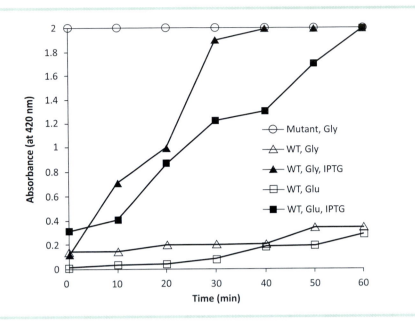

Figure A2.21 Final form of XY scatter chart after formatting in Excel. Additional changes can be made in a Word document (2007 or later) after copying and pasting the graph from Excel.

ship between the two variables, in which case it doesn't matter which variable is plotted on which axis.

A good way to see the relationship between two quantitative variables is to plot the measured data as a **scatterplot**. On a scatterplot, *the data points are not connected with lines*. The idea is to look for a trend that may ultimately be expressed as a mathematical function. This so-called trendline or regression line or "best-fit" line can then be used to make predictions about one variable when the other is known. How well the trendline fits the measured data is described by the R-squared (R^2) value. The closer the R^2 value is to 1, the more reliable the trendline.

Excel makes it possible to add six different types of trendlines to a scatterplot. The type of trendline you use depends on the pattern of the data points and, equally importantly, the theoretical basis of the physical phenomenon. For example, Beer's Law states that the absorbance of a solution is proportional to its concentration. Thus the correlation between these two variables should be linear (Figure A2.22A). On the other hand, studies on enzyme kinetics have shown that the initial velocity of an enzymatic reaction increases logarithmically with substrate concentration (Figure A2.22B).

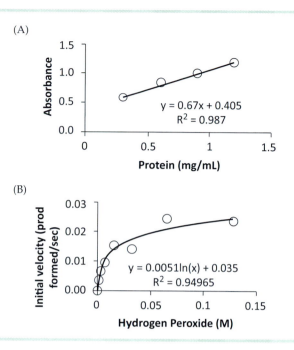

Figure A2.22   Trendlines added to scatterplots show the correlation between the variables. (A) Standard curve for Bradford protein assay. A linear trendline is used because, according to Beer's Law, absorbance is proportional to concentration. (B) Michaelis-Menten curve for an enzyme-substrate reaction. The logarithmic trendline is based on the principles of enzyme kinetics, whereby initial velocity levels off when all enzymes are saturated with substrate. Trendline equations make it possible to predict one variable when the other is known. The R-squared values close to 1 indicate that the trendlines fit the measured data very well.

It would be incorrect to add a linear trendline to an enzyme curve or a logarithmic trendline to an absorbance curve. When using trendlines, therefore, make sure you understand the theoretical basis of the system.

Linear trendline—A standard curve for a protein assay

A standard curve is a common example in biology where a linear trendline and its equation are used to predict one variable when the other is known. For example, standard curves are often used to:

- Predict the protein concentration of an unknown sample using the correlation between absorbance and known protein concentration (e.g., biuret method, Bradford method)

- Determine the size of a DNA fragment by comparing the distance it migrated on a gel to the distance migrated by DNA fragments whose sizes are known (the molecular-weight marker)
- Quantify nitrogen or phosphorous in aquatic ecosystems.

A standard curve for a protein assay involves preparing a series of solutions with known protein concentration, adding a reagent to the protein solutions, which binds to specific amino acid residues or peptide bonds, and then measuring the absorbance of each sample. Because of Beer's Law, as protein concentration increases, absorbance increases in a linear manner within the sensitivity range of the assay. The standard curve consists of a graph with absorbance on the *y*-axis and protein concentration on the *x*-axis.

Entering data in worksheet. Type *x*-axis data (protein concentrations) in Column A and *y*-axis data (absorbance values) in Column B.

Creating the scatter chart. Select the data in Columns A and B.

1. In Excel 2010, click **Insert | Charts | Scatter | Scatter with only Markers** (Figure A2.23). In Excel for Mac 2011, the command sequence is **Charts | Scatter | Marked Scatter**.
2. Follow the formatting instructions for XY graphs, Steps 3–12 on pp. 220–232.
3. To insert a trendline in Excel 2010, click anywhere in the plot or chart area and then click **Chart Tools Layout | Analysis |**

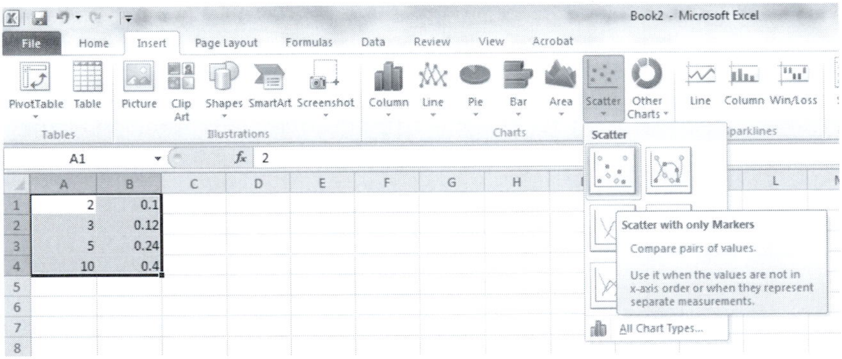

Figure A2.23 In Excel 2010, choose **Scatter with only Markers** if you plan to add a trendline later. In Excel for Mac 2011, choose **Charts | Scatter | Marked Scatter**.

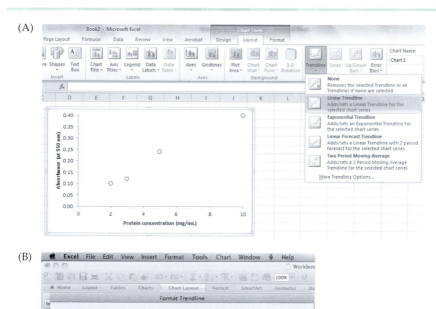

Figure A2.24   Insert a linear trendline by clicking the graph. (A) In Excel 2010, select **Chart Tools Layout | Analysis | Trendline | Linear Trendline**. (B) In Excel for Mac 2011, select **Chart Layout | Analysis | Trendline** and then select **Linear** under **Type | Trend/Regression type**.

Trendline | Linear Trendline (Figure A2.24A). In Excel for Mac 2011, click the plot or chart area and then **Chart Layout | Trendline | Type: Linear** (Figure A2.24B).

4. If you plan to insert a trendline *and* display the equation and R-squared value on the chart in Excel 2010, right-click a marker and select **Add Trendline**.

5. In the **Format Trendline** dialog box (Figure A2.25), under Trendline Options, click **Linear, Display Equation on chart**, and **Display R-squared value on chart**. Then click **Close**. The equation allows you to predict the concentration in an

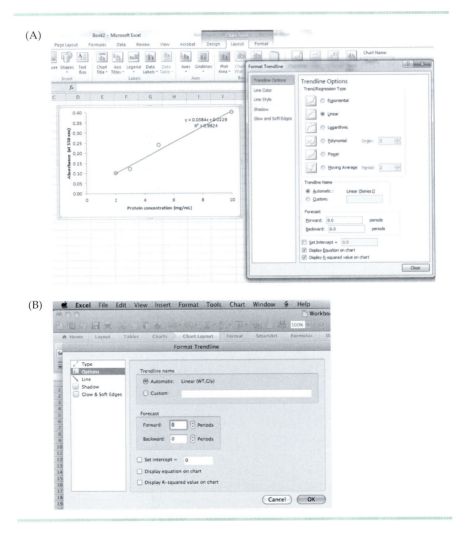

Figure A2.25 Display the equation and R-squared value of the trendline by right-clicking one of the markers. (A) In Excel 2010, click the **Display Equation on chart** and **Display R-squared value on chart** checkboxes. (B) In Excel for Mac 2011, select **Add Trendline** to open the **Format Trendline** dialog box. Under **Options**, click the appropriate checkboxes.

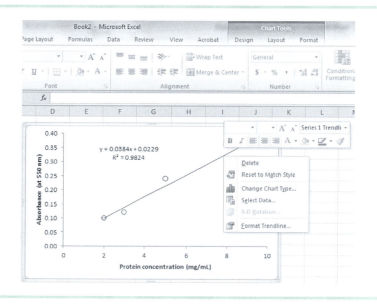

Figure A2.26 To format an already existing trendline in both Excel 2010 and Excel for Mac 2011, right-click the trendline (the drop-down menu will be displayed) and click **Format Trendline**. Or, to skip the drop-down menu step, double-click the trendline to open the **Format Trendline** dialog box directly.

"unknown" sample after measuring its absorbance. The R-squared value tells you how well the equation fits the data.

6. Drag the text box containing the equation and the R-squared value away from the trendline to an open space in the plot area.

7. Protein assays typically have a so-called sensitivity range, a range of concentrations for which the assay is reliable. For the biuret assay, this range is 1 to 10 mg/mL. To extrapolate the trendline backwards to 1 mg/mL to encompass the lower limit of the assay, right-click the trendline to display a drop-down menu (Figure A2.26). Click **Format Trendline** and then enter the number "1" in the box for **Forecast | Backward** (see Figure A2.25). Then click **Close**. The number you enter in the box extends the line backward by that number of units from the original origin of the line.

The final graph, formatted according to CSE Manual recommendations, looks like Figure A2.27. Further changes to the graph can be made in Word 2010 after copying it from Excel 2010. Because XY graphs with trendlines are very common in biology, consider saving the format in a chart template (see pp. 257–260).

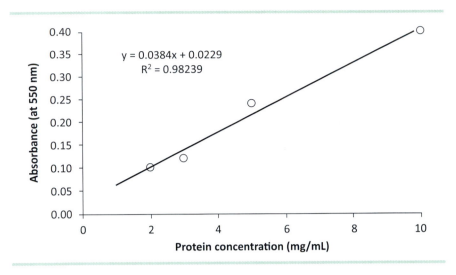

Figure A2.27 Final form of XY scatter chart with a trendline after formatting in Excel. Additional changes can be made in a Word document (2007 or later) after copying and pasting the graph from Excel.

Plotting Bar Graphs

Bar graphs are used to compare individual data sets when one of the variables is categorical (not quantitative).

Column charts

Column charts are bar graphs with vertical bars.

ENTERING DATA IN WORKSHEET

Before you enter data in an Excel worksheet, you must have a clear idea of what your column chart should look like. By convention, the feature that all the columns have in common (the variable that was measured) lies on the axis parallel to the columns. Enter the values for this feature in Column B and the labels for the categories in Column A (Figure A2.28). The categories should be sequential, with the control treatment column farthest left. If there is no particular order to the categories, arranging the bars from shortest to longest (or vice versa) makes the results easier to comprehend. When deciding what order to enter the categories in the worksheet, remember that the *lowest* row number contains the category label for the *leftmost* column.

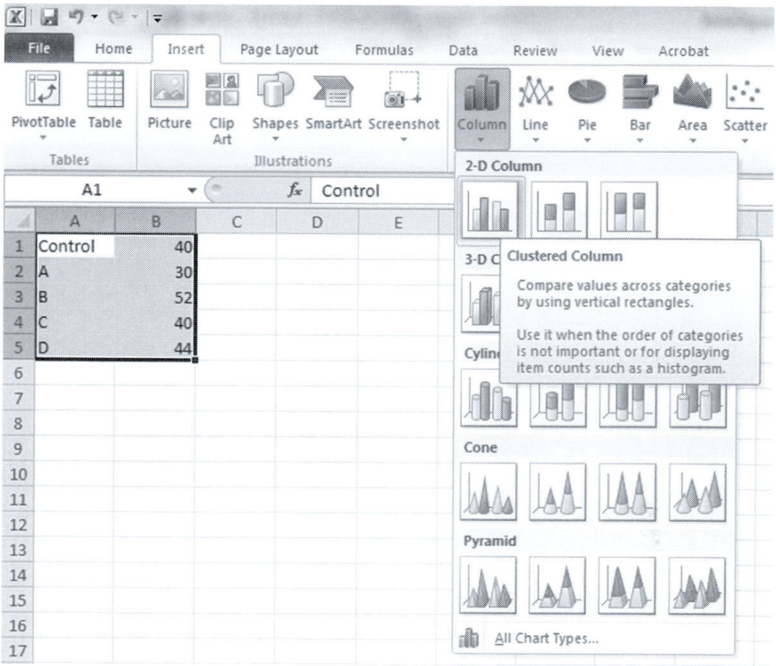

Figure A2.28   The first step in making a column chart (vertical bar graph) in Excel 2010 is to click **Insert | Charts | Column | 2-D Column, Clustered Column**. In Excel for Mac 2011, the command sequence starts on the **Charts** tab.

CREATING THE COLUMN CHART

1. Select the data you just entered in Columns A and B. In Excel 2010, click **Insert | Charts | Column | 2-D Column, Clustered Column** (see Figure A2.28). In Excel for Mac 2011, the command sequence starts on the **Charts** tab.

2. Follow the instructions for plotting XY graphs, Steps 3–6 and 12, as described on pp. 220–232.

3. All of the columns in the graph should be the same width and the columns should always be wider than the space between them. To adjust the width of the columns, double-click or right-click one of them. In Excel 2010, in the **Format Data Series** dialog box (Figure A2.29), drag the slider for **Gap Width** to the left to decrease the gap and simultaneously increase the width of

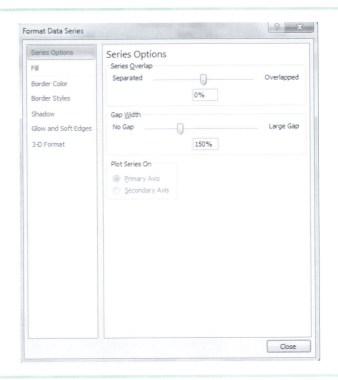

Figure A2.29   Double-clicking a column opens the **Format Data Series** dialog box. Decrease the gap width (while simultaneously increasing the column width) by dragging the **Gap Width** slider to the left. In Excel for Mac 2011, click the down arrow to decrease the **Gap Width** number.

the columns. In Excel for Mac 2011, click the down arrow to decrease the **Gap Width** number.

4. Still in the **Format Data Series** dialog box, click **Fill | Solid fill | Fill Color: Black** to make the columns black against a white background for best contrast in black-and-white publications. The procedure is the same for Excel 2010 and Excel for Mac 2011.

Older versions of Excel allowed you to apply a black-and-white pattern fill to the columns. Although this function has been removed from Excel 2010, you can install an add-in called PatternUI. Type "excel patterns" into the Google search box and click the "Microsoft Excel: Chart Pattern Fills" link to access this add-in (http://blogs.msdn.com/excel/archive/2007/11/16/chart-pattern-fills.aspx [accessed 2012 Jul 4]). Once you have downloaded the add-in to your computer, install it by clicking **File | Options | Add-ins**. Click the **Go...** button at the bottom of the dialog box. Then click **Browse...** and navigate to the location where you saved the PatternUI file. Click it and then select **OK**. Make sure there is a checkmark next to this file in the **Add-ins** dialog box and click **OK**. When you open a new or existing Excel 2010 workbook, a new **Patterns** button will be available on the **Chart Tools Format** tab. To apply a pattern, click a column and then make a selection from the **Patterns** drop-down menu.

5. To remove the border around the graph, double-click anywhere in the chart area or right-click and select **Format Chart Area**. Inside the **Format Chart Area** dialog box, click **Border Color | No line** and then **Close**. In Excel for Mac 2011, inside the dialog box click **Line | No Line**.

6. Depending on the discipline (or your instructor's preference), the baseline may or may not be visible, but all the columns must be aligned as if there were a baseline. To remove the baseline, double-click a category title on the *x*-axis to open the **Format Axis** dialog box. Click **Line Color | No line** and then click **Close**. The procedure is the same for Excel 2010 and Excel for Mac 2011.

The final graph, formatted according to CSE Manual recommendations, looks like Figure A2.30. Further changes to the graph can be made in Word

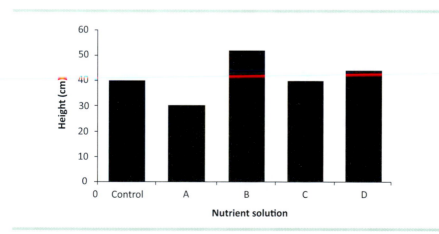

Figure A2.30   Final form of column chart after formatting in Excel. Additional changes can be made in a Word document (2007 or later) after copying and pasting the graph from Excel.

2010 after copying it from Excel 2010. Because column charts are common in some disciplines, consider saving the format in a chart template (see pp. 257–260).

Clustered column charts

Clustered columns may represent the results of different treatments after the same period of time or the results of the same treatments after different periods of time. Each column in the cluster must be easy to distinguish from its neighbor. Colorful columns may look good on your computer screen, but they may turn out to be the same shade of gray when the graph is printed on a black-and-white printer. If the person who evaluates your work receives a black-and-white copy of your paper, be sure to proofread the hardcopy and check that it's clear which treatments the columns represent.

ENTERING DATA IN WORKSHEET

Before you enter data in an Excel worksheet, you must have a clear idea of what your column chart should look like. The category labels are entered in Column A of the worksheet and should be sequential. The measured values for the columns in each cluster are entered in Column B, C, and so on. Enter a short title in the first row of Columns B, C, and so on, so that Excel can use these titles to generate the legend. Do not enter a title in Column A, because category labels are not part of the legend.

CREATING THE CLUSTERED COLUMN CHART

1. Select the data you just entered in Columns A, B, and C, including the titles in the first row. In Excel 2010, click **Insert | Charts | Column | 2-D Column, Clustered Column** (see Figure A2.28). In Excel for Mac 2011, the command sequence starts on the **Charts** tab.
2. Follow the instructions for plotting column charts on pp. 242–246.

The final graph, formatted according to CSE Manual recommendations, looks like Figure A2.31. Further changes to the graph can be made in Word 2010 after copying it from Excel 2010. Because clustered column charts are common in some disciplines, consider saving the format in a chart template (see pp. 257–260).

Bar charts

Bar charts are bar graphs with horizontal bars. Bar charts are more practical than column charts when the category labels are long.

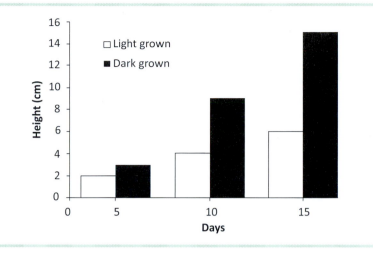

Figure A2.31   Final form of a clustered column chart after formatting in Excel. Additional changes can be made in a Word document (2007 or later) after copying and pasting the graph from Excel.

ENTERING DATA IN WORKSHEET

Before you enter data in an Excel worksheet, you must have a clear idea of what your bar chart should look like. By convention, the feature that all the bars have in common (the variable that was measured) lies on the axis parallel to the bars. Enter the values for this feature in Column B and the labels for the categories in Column A (Figure A2.32). The categories should be sequential, with the control treatment bar on top (or on the bottom). If there is no particular order to the categories, arranging the bars from short-est to longest (or vice versa) makes the results easier to comprehend. When deciding in what order to enter the categories in the worksheet, remember that the *lowest* row number contains the category label for the *lowest* bar.

CREATING THE BAR CHART

1. Select the data you just entered in Columns A and B. In Excel 2010, click **Insert | Charts | Bar | 2-D Column, Clustered Bar** (see Figure A2.32). In Excel for Mac 2011, the command sequence starts on the **Charts** tab.

2. Follow the instructions for plotting column charts on pp.242–246.

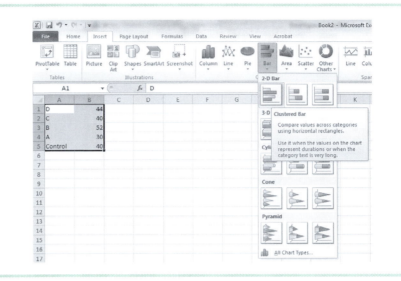

Figure A2.32 The first step in making a bar chart (horizontal bar graph) is to click **Insert | Charts | Bar | 2-D Column, Clustered Bar**. In Excel for Mac 2011, the command sequence starts on the **Charts** tab.

The final graph, formatted according to CSE Manual recommendations, looks like Figure A2.33. Further changes to the graph can be made in Word 2010 after copying it from Excel 2010. Because bar charts are common in some disciplines, consider saving the format in a chart template (see pp. 257–260).

Pie Graphs

Pie graphs are commonly used to show financial data, but they are seldom used in biology research papers. The CSE Manual recommends a table rather than a pie graph for showing percentage of constituents out of the whole. If you would like to make a pie graph, however, follow these instructions.

Entering data in worksheet

Let's say you did a survey of insects found in your backyard. Enter the name of each kind of insect in Column A and sort them according to percentage in Column B (Figure A2.34). In other words, put the most abundant insect in the first row, the second most abundant one in the second row, and so on. This arrangement will result in the largest segment of the pie beginning at 12 o'clock, with the segments decreasing in size clockwise. There should be

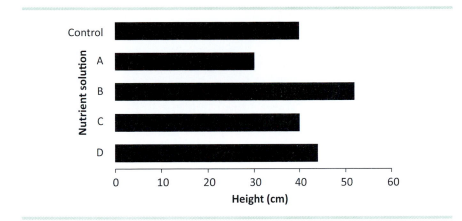

Figure A2.33   Final form of bar chart after formatting in Excel. Additional changes can be made in a Word document (2007 or later) after copying and pasting the graph from Excel.

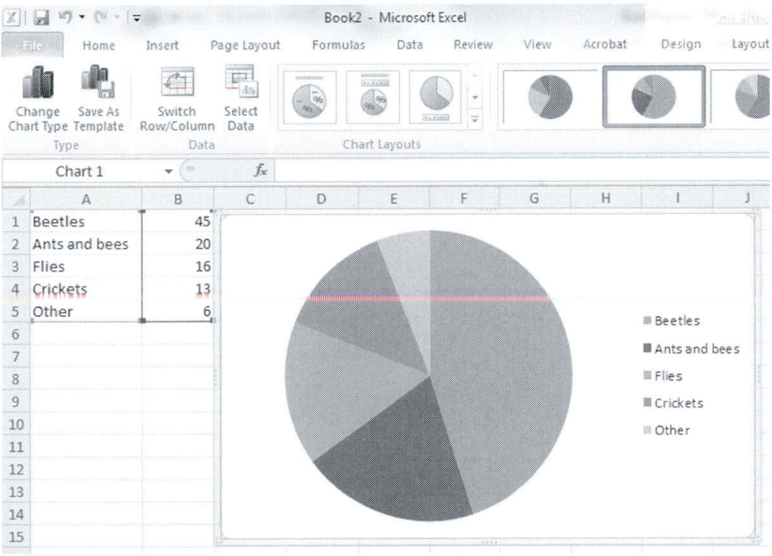

Figure A2.34   The first step in making a pie chart is to click **Insert | Charts | Pie**. In Excel for Mac 2011, the command sequence starts on the **Charts** tab. Order the categories according to percentage (highest to lowest) to plot the largest segment starting at 12 o'clock.

between two and eight segments in the pie. Combine small segments (those less than 5%) under the heading "Other."

Creating the pie chart

1. Select the data you just entered in Columns A and B. In Excel 2010, click **Insert | Charts | Pie | 2-D Pie**. In Excel for Mac 2011, the command sequence starts on the **Charts** tab.

2. Click the plot area or chart area of the newly created chart to activate the **Chart Tools** contextual tab, which has three tabs of its own. On the **Design** tab, click **Chart Layouts | Layout 1** (Figure A2.35A). In Excel for Mac 2011, click the chart and then

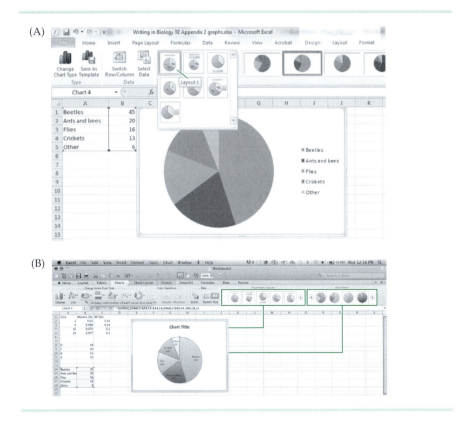

Figure A2.35 (A) In Excel 2010, click the chart and then select the first **Chart Layout** and a suitable **Chart Style** from the **Design** tab. (B) In Excel for Mac 2011, click the chart and then select a suitable **Chart Quick Layout** and **Chart Style** on the **Charts** tab.

select a suitable **Chart Quick Layout** and **Chart Style** on the **Charts** tab (Figure A2.35B).

3. Delete the Chart Title text box unless you are preparing this graph for an oral presentation.

4. According to CSE guidelines, the labels for each segment should be outside of the pie. In Excel 2010, click **Chart Tools Layout | Labels | Data Labels | Outside End** to position the labels on the outside. The procedure is the same for Excel for Mac 2011.

5. To outline the pie pieces to make them easier to distinguish, right-click the pie chart and select **Format Data Series**. Select **Border Color: Solid Line | White** and **Border Styles | Width: 2 pt**.

6. To remove the border around the graph, double-click anywhere in the chart area or right-click and select **Format Chart Area**. Inside the **Format Chart Area** dialog box, click **Border Color | No line** and then **Close**. In Excel for Mac 2011, inside the dialog box click **Line | No Line**.

The final graph, formatted according to CSE Manual recommendations, looks like Figure A2.36. Further changes to the graph can be made in Word 2010 after copying it from Excel 2010. For example, if you notice a mistake in the percentages, right-click the plot area or chart area and select **Edit Data**.

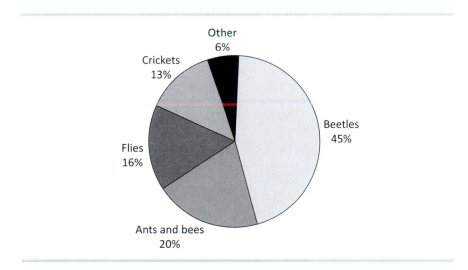

Figure A2.36   Final form of pie chart after formatting in Excel. Additional changes can be made in a Word document (2007 or later) after copying and pasting the graph from Excel.

This command links you directly to the Excel worksheet and any changes you make in the worksheet are automatically transferred to the graph in the Word 2010 document.

Error Bars and Variability

The reliability of scientific data depends on good experimental design, the skill and experience of the person collecting the data, the reliability of the equipment, and the proven effectiveness of the methods and procedures, among other things. Even when all of these factors have been optimized, there is still a strong likelihood that there will be variability in the measurement data. For example, different students measuring the same sample with the same equipment may come up with different results. Seasoned scientists measuring the same variable in replicate experiments are likely to find slight variations in their data. Genetically identical seeds from the same lot, planted in the same soil and watered at the same time, may germinate at different times and grow at different rates. How can we be confident that our results accurately represent the phenomena we are trying to understand when there is variability in the measurement data?

One way to depict variability is to show all of the measured data on a scatterplot, as in Figure A2.37A. This approach, however, makes it hard to see if there is a trend or relationship between the dependent and independent variables. To reduce the amount of data and begin to make sense of the values, we can take the average of multiple measurements as our best estimate of the true value. In statistics, the average is called the arithmetic mean and it is calculated by dividing the sum of all the values by the number of values. The formula for calculating the mean value in Excel is

=AVERAGE(…),

where "…" is the range of cells to be averaged.

A graph of the mean values is less cluttered (Figure A2.37B), but potentially important information about the variability has been lost. There are two common statistical methods for describing variability: standard deviation and standard error of the mean. Both measures are based on a statistic called variance, which describes how far each measurement value is from the mean. Standard deviation is the square root of the variance and standard error is the standard deviation divided by the square root of the number of measurements. Your handheld calculator and Excel both make it easy to calculate standard deviation and standard error, but "plugging and chugging" is not the same as understanding what you're doing. It is worth your while to take a statistics course to learn how these formulas are derived and how to use and interpret statistics appropriately.

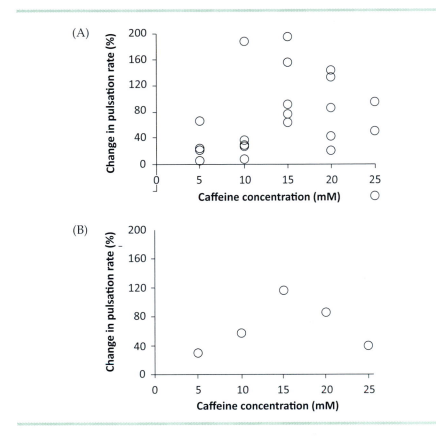

Figure A2.37   (A) All measured data plotted as a scatter graph, and (B) Mean values for effect of caffeine on change in pulsation rate of blackworms. Mean values reduce the amount of data but they do not reflect the degree of variability in the raw values.

Adding error bars about the means

Standard deviation and standard error can be depicted graphically in the form of error bars around the means. To add error bars to the mean values, follow these steps:

1. Select an empty cell to enter the formula for standard deviation, which is

=STDEV(…)

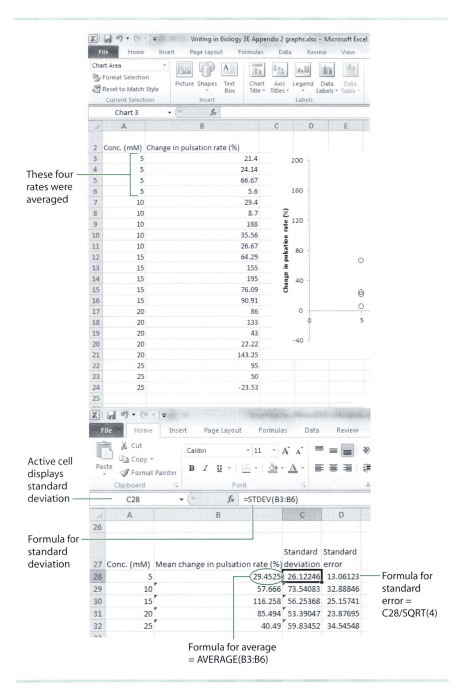

These four rates were averaged

Active cell displays standard deviation

Formula for standard deviation

Formula for standard error = C28/SQRT(4)

Formula for average = AVERAGE(B3:B6)

Figure A2.38 Calculation of mean, standard deviation, and standard error.

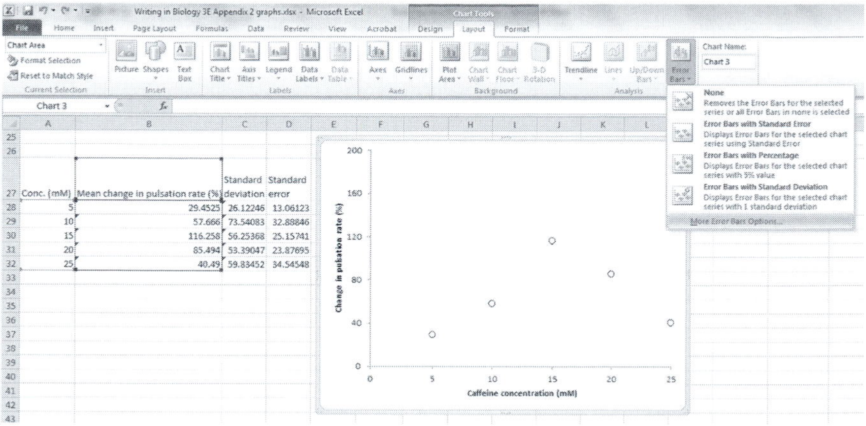

Figure A2.39   The first step in adding error bars to a scatter plot is to click **Chart Tools Layout | Analysis | Error Bars**.

where "…" is the range of cells that were averaged to calculate the mean. As shown in Figure A2.38, the four rates for the 5 mM caffeine solution were averaged using the formula "=AVERAGE(B3:B6)" and the mean value is given in cell B28.

2. Select another empty cell to enter the formula for standard error, which is

STDEV/√N

or, in Excel format,=Cell reference for STDEV/SQRT(N)

In this example, C28 is the cell that contains the value for standard deviation, SQRT is the formula for square root, and N is the number of measurement values that were averaged to calculate the mean.

3. Now click the chart area to display the **Chart Tools Layout** tab. In the **Analysis** group, click **Error Bars | More Error Bars Options** (Figure A2.39). This opens the **Format Error Bars** dialog box (Figure A2.40). If you would like the vertical error bars to extend above and below the mean value, keep the default **Direction: Both**. Under **Error Amount**, click **Custom** and then **Specify Value** to open a second small **Custom Error Bars** dialog box. Click the red arrow for **Positive Error Value** and select the cells in your worksheet that contain the standard error values.

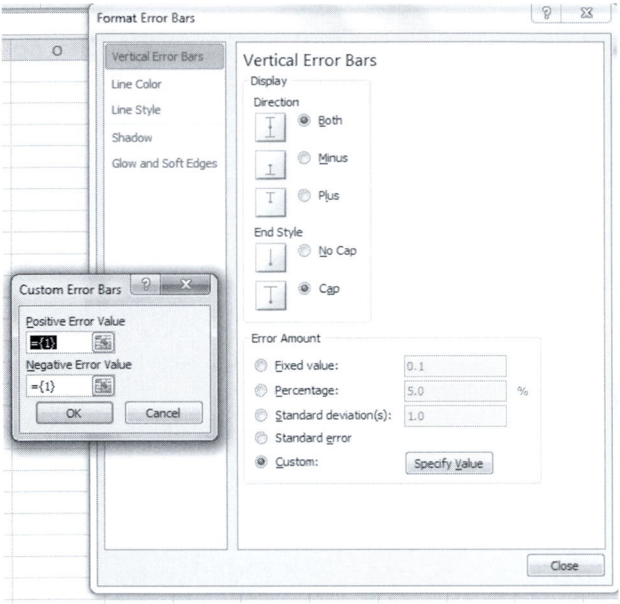

Figure A2.40   Selection of **Error Amount: Custom** allows you to enter the range of cells in your worksheet that contains the standard error.

> Click the red arrow again to close the box. Repeat this process for **Negative Error Value**. Click **OK** to close the dialog box. Finally, click **Close** to close the **Format Error Bars** dialog box.

If your version of Excel for some crazy reason has inserted horizontal error bars, delete them by clicking any one of them and pressing **Delete**. Your graph of the mean values with error bars will now look like Figure A2.41.

Data analysis with error bars

The addition of error bars to the mean values changes our interpretation of the results. The larger the standard error, the less confidence we have that the mean represents the true value. Furthermore, the more the error bars overlap, the less likely that these measurement values differ significantly from each other. Based on the mean values alone, we might have concluded that there was a real difference in pulsation rate with every 5 mM increase in caffeine concentration (see Figure A2.37B). The large overlap between the error bars for 5 and 10 mM caffeine, however, suggests that the pulsation rates do not differ that much for these concentrations (see Figure A2.41). On

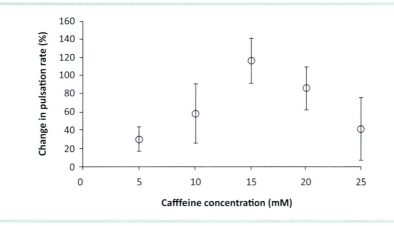

Figure A2.41   Vertical error bars inserted about the mean value to show variance.

the other hand, the error bars for 10 and 15 mM caffeine barely overlap, so we are more confident that there may be a real difference in pulsation rate between these two concentrations.

Making Chart Templates

When you have formatted XY graphs, bar graphs, and pie graphs exactly the way you want, you can save the format as a chart template (*crtx). The next time you need to make a similar graph, you can select the chart template and apply it to your data. Instructions for making a chart template for an XY graph with one data set are given below, but the procedure is the same for other types of graphs.

1. Select an XY graph that you have formatted according to the instructions on pp. 220–232. Omit Steps 7–9. In other words, keep all the Axis Options (Minimum, Maximum, Major Unit and Minor Unit) on **Auto**.

2. In Excel 2010, click the chart area to display the **Chart Tools Layout** tab. Under **Properties | Chart Name:** type "Scatter with Straight Lines". In Excel for Mac 2011, replace the chart number with "Straight Marked Scatter".

3. On the **Chart Tools Design** tab, click **Type | Save as Template** and type "Scatter with Straight Lines" in the **File name** box (Figure A2.42A). Notice that the **Save as type** is "Chart Template Files (*crtx)". In Excel for Mac 2011, the command sequence is **Charts | Other | Save as Template** (Figure A2.42B).

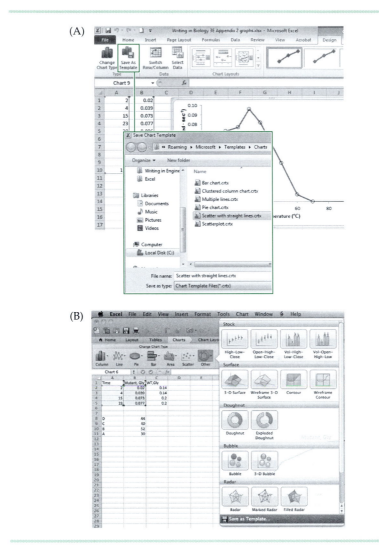

Figure A2.42 To save the formatting of a graph, click the graph that you've formatted correctly. (A) In Excel 2010, click **Chart Tools Design | Type | Save as Template** and give the template a name. (B) In Excel for Mac 2011, click the **Charts** tab and select **Other** among the chart types. Click **Save as Template** at the bottom of the dialog box and give the template a name.

4. To apply this format to new data, enter the data as described on p. 220. In Excel 2010, select the data and click **Insert | Charts** diagonal arrow button ⬒. In the **Insert Chart** dialog box, click

Templates on the left panel (Figure A2.43A). Holding the mouse over each chart template opens a pop-up window with a description of that template. Click **My Templates | Scatter with Straight Lines** and then **OK**. In Excel for Mac 2011, on the

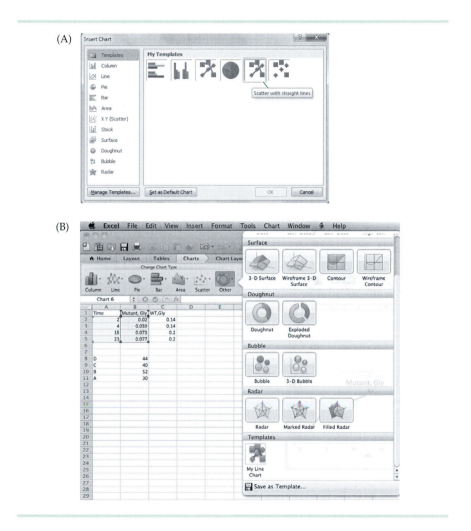

Figure A2.43   To use one of the templates you've saved, select the data to be graphed. (A) In Excel 2010, launch the **Insert Chart** dialog box by clicking the **Insert | Charts** diagonal arrow button. On the left panel, click **Templates** to display the chart templates that you have saved. The name of each template is shown when holding the mouse over the individual icons. Select the appropriate template. (B) In Excel for Mac 2011, on the **Charts** tab, click **Other | Templates** and select the desired template.

Charts tab, click **Other | Templates** and select the desired template (Figure A2.43B).

5. The graph will most likely not be formatted correctly. To correct the problem in Excel 2010, click **Chart Tools Design | Data | Switch Row/Column** twice. In Excel for Mac 2011, click **Charts | Switch Plot** and then click **Plot series by row** followed by **Plot series by column**.

6. Click each **Axis Title** text box and type the desired title.

7. If necessary, adjust the scale of each axis as explained in Steps 7–9 on pp. 226–229.

PREPARING ORAL PRESENTATIONS WITH MICROSOFT POWERPOINT 2010

Introduction

This appendix deals exclusively with Microsoft PowerPoint 2010 using the Windows 7 operating system. While there are similarities between the Power-Point 2010 and PowerPoint for Mac 2011 Ribbon interfaces, some of the commands are accessed differently than shown here.

Microsoft PowerPoint allows you to create a slide show containing text, graphics, and even animations. If you have been using Office 2007, then Office 2010 will not seem that different. However, if you are a Mac user, the Ribbon at the top of the screen is a new addition to the user interface. The **Ribbon** is a single strip that displays **commands** in task-oriented **groups** on a series of **tabs** (Figure A3.1). The new **File** tab replaces the Microsoft Office Button that was located in the top left of the screen in PowerPoint 2007. Additional commands in some of the groups can be accessed with the **Dialog Box Launcher**, a diagonal arrow in the right corner of the group label. The **Quick Access Toolbar** comes with buttons for saving your file and undoing and redoing commands; you can also add buttons for tasks you perform frequently.

The window size and screen resolution affect what you see on the Ribbon. You may see fewer command buttons or an entire group abbreviated to one button on a smaller screen or if you have set your screen display to a lower resolution. If you want to see the tab names without the command buttons, click **Minimize the Ribbon**.

Clicking the **File** tab opens **Backstage View** (Figure A3.2). The left panel lists the file-related commands, the center panel provides the options for

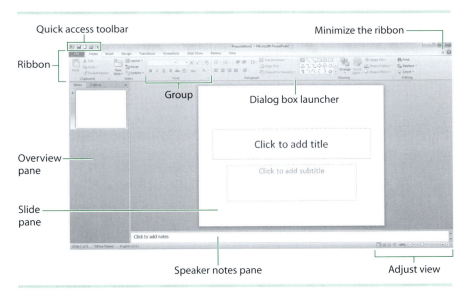

Quick access toolbar

Minimize the ribbon

Ribbon

Group

Dialog box launcher

Click to add title

Click to add subtitle

Overview pane

Slide pane

Click to add notes

Speaker notes pane

Adjust view

Figure A3.1 Screen shot of a PowerPoint 2010 presentation in **Normal** view. The command buttons are located on Ribbon tabs at the top of the screen. View options are on the status bar in the lower right corner.

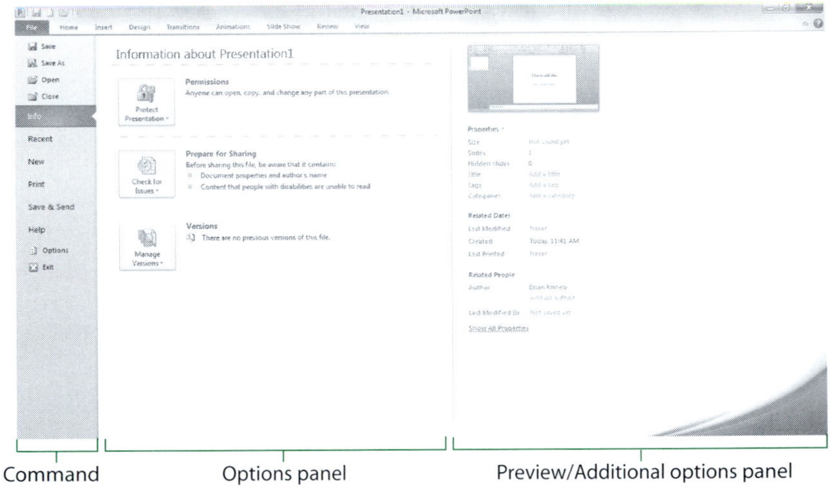

Information about Presentation1

Command

Options panel

Preview/Additional options panel

Figure A3.2 Backstage View is displayed when clicking the **File** tab. The left panel shows file-related commands such as Save, Save As, Open, Close, and Print. The center and right panels show options related to the command selected on the left panel.

that command, and the right panel shows a preview or other options. Document commands such as Save, Save As, Open, Close, Recent, New, and Print, which were located on the **Office Button** in PowerPoint 2007 and on the **File** menu in PowerPoint 2003, are located on the **File** tab. Other commands on the **File** tab are

- **Info**, which provides new file management tools for setting permissions, inspecting the document for hidden properties and personal information, and recovering earlier autosaved or unsaved versions of a file.

- **Save & Send** makes it easier to collaborate with others on files. All collaborators must use Office 2010 and possess SharePoint 2010 software or a SkyDrive account to access and save the files online. Similar to Google docs, this feature allows two or more people to work on the same document at the same time.

- **Help.** The **Help** menu can also be accessed by pressing the **F1** function key.

- **Options** allows you to change program settings such as user name and initials, rules for checking spelling and grammar, AutoCorrect options, how often documents are saved automatically, and so on. The **Options** button is also the gateway to commands for customizing the Ribbon and the Quick Access Toolbar and for managing Microsoft Office add-ins and security settings.

AutoCorrect. Unformatted AutoCorrect entries programmed in Word also work in PowerPoint, but entries containing symbols or italics do not (see the section on "AutoCorrect" in Appendix 1).

Customize the Ribbon. This feature lets you create a new tab on the Ribbon, to which you can add hard-to-find commands that you use frequently. Click **File | Options | Customize Ribbon**. After clicking the **New Tab** button, add groups and commands to the new tab (see Figure A1.6).

Handling Computer Files

The section "Good Housekeeping" in Appendix 1 applies equally to Word documents and PowerPoint presentations. Read over this section to develop good habits for naming, organizing, and backing up computer files. In addition, if you are making the transition from Microsoft Office 2003 (or earlier) to 2010, there are some things you should know about file compatibility, which is also covered under "Good Housekeeping."

Commands in PowerPoint 2010

The best way to get up to speed with PowerPoint 2010 is to learn where the commands you used most frequently in earlier versions of PowerPoint are located. Table A3.1 is designed to help you with this task. Frequently used commands are listed in alphabetical order in the first column and the PowerPoint 2010 command sequence is given in the second column. The PC keyboard shortcuts for most of the commands are shown in the third column. If you are a Mac user,

Ctrl = Command (⌘)

Alt = Option (however, some Alt key sequences do not have a
        Mac equivalent)

The nomenclature for the command sequence is as follows: **Ribbon tab | Group | Command button | Additional Commands** (if available). For example, **Home | Slides | New Slide ▼ | Title Only** means "Click the **Home** tab and in the **Slides** group, click the down arrow (▼) on the **New Slide** button to select the **Title Only** option."

Microsoft Office also has online training materials to help you transition from earlier versions of PowerPoint to PowerPoint 2010: Go to Microsoft's Home Page (http://office.microsoft.com) and type "getting started with powerpoint 2010" into the search box. Choose your situation and follow the instructions on the website.

Designs for New Presentations

To start your PowerPoint presentation, click **Start | All Programs | Microsoft PowerPoint 2010** and then **File | New** to open the gallery of available templates and themes (Figure A3.3). You can start with a Blank Presentation, select a Sample template or Theme, use the format of an existing presentation, or download a template or theme from Microsoft Office Online. Designs with a light background make the room much brighter during the actual presentation, which has the dual advantage of keeping the audience awake and allowing you to see your listener's faces. Many professional speakers, however, prefer white text on a dark background because the text appears larger. Whatever your preference, choose a design that reflects your style, is appropriate for the topic, and complements the content of the individual slides.

When you choose one of the Themes (called "design templates" in earlier versions of PowerPoint), the slides in your presentation will have a consistent and professional appearance. The color schemes for the background, text, bullets, and illustrations (shapes and SmartArt) were designed to be

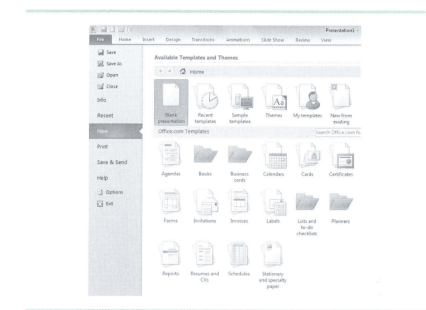

Figure A3.3   Select themes and templates for new presentations by clicking **File | New**.

TABLE A3.1	Common commands (listed alphabetically) and how to carry them out in PowerPoint.	
Command	PowerPoint 2010 Command Sequence[a]	PC Hot Key
Editing		
See Table A1.3		
File		
See Table A1.3		
Formatting		
See Table A1.3		
Handouts		
Headers and Footers	Insert	Text	Header & Footer	Notes and Handouts tab	Alt+NH

[a]Many of the Word 2010 commands can also be accessed by selecting text or an object and right-clicking. For Alt commands, hold down **Alt** while pressing the first letter key, which corresponds to a tab name [F][H][N][P][S][M][R][W][X][B]. Then release **Alt** and type the next letter(s) in the sequence. F11 is a function key.

TABLE A3.1   *Continued*

Command	PowerPoint 2010 Command Sequence[a]	PC Hot Key
Master	View | Master Views | Handout Master	Alt+WH
Print	File | Print | Full Page Slides | Handouts	Alt+PH
Print		
Handouts	File | Print | Full Page Slides | Handouts	Alt+FPH
Slides	File | Print | Full Page Slides	Alt+FPH
Speaker notes	File | Print | Full Page Slides | Notes Pages	Alt+FPH
Slides and notes economically	File | Save & Send | Create Handouts	Alt+FDH
Slide Show		
End slide show	Right-click + End	Esc
Go to the next slide	Left-click	Enter or space bar
Go to the previous slide	Right-click + Previous	Backspace
Mouse pointer		
As a laser pointer		Ctrl+left mouse button
As a pen or highlighter	Right-click | Pointer Options | Pen (Highlighter)	Right-click + OP (OH)
Start at the beginning	Slide Show | Start Slide Show | From Beginning	F5 or Alt+SB
Start at the current slide	Slide Show | Start Slide Show | From Current Slide	Shift+F5 or Alt+SC
Slides		
Animations	Animations tab	Alt+A
Background	Design | Background	Alt+GB (GG)
Copy	Click a slide on the overview pane or in Slide Sorter view. Home | Clipboard | Copy	Ctrl+C or Alt HC
Delete	Click a slide on the overview pane or in Slide Sorter view and press Backspace or Delete	Alt+WL or Alt+WI, select slide, Backspace or Delete
Duplicate	Home | Slides | New Slide ▼ | Duplicate Selected Slides	Ctrl+V or Alt+HID

TABLE A3.1 *Continued*		
Command	**PowerPoint 2010 Command Sequence[a]**	**PC Hot Key**
Headers and Footers	Insert | Text | Header & Footer | Slide tab	Alt+NH
Insert new	Home | Slides | New Slide	Ctrl+M or Alt+HI
Layout	Home | Slides | Layout	Alt+HL
Print	File | Print | Full Page Slides	Ctrl+P or Alt+P
Slide Master	View | Master Views | Slide Master	Alt+WM
Templates, design	See Themes	
Themes		
Start with new blank	File | New | Blank presentation	Alt+FNL
Start with built-in theme	File | New | Themes	Alt+FNI
Retrofit existing presentation	Design | Themes	Alt+GH
Transitions	Transitions tab	Alt+K
View		
Normal	View | Presentation Views | Normal	Alt+WL
Slide Sorter	View | Presentation Views | Slide Sorter	Alt+WI
Slide Show	Slide Show | Start Slide Show |	
Start from beginning	From Beginning	F5 or Alt+SB
From current slide	From Current Slide	Shift+F5 or Alt+SC
Visual elements		
Charts (graphs)	Copy and paste from Excel	
Hyperlinks, insert	First click object or text	
Action button	Insert | Links | Action	Alt+NK
Insert Hyperlink button	Insert | Links | Hyperlink	Ctrl+K or Alt+NI
Line drawings	See Shapes	
Media, insert	Insert | Media | Video (Audio) or Home | Slides | New Slide ▼ | Title and Content | Insert Media Clip icon	Alt+NV (NO)

TABLE A3.1 *Continued*		
Command	**PowerPoint 2010 Command Sequence**[a]	**PC Hot Key**
Pictures	Insert	Images	Pictures	Alt+NP
Shapes	Insert	Illustrations	Shapes	Alt+NSH
SmartArt Graphics	Insert	Illustrations	SmartArt or Insert	Home	Slides	New Slide ▼	Title and Content	Insert SmartArt Graphic icon	Alt+NM or Alt+HI
Tables		
Format	Click inside table. Table Tools Design and Layout tabs	Alt+JT and Alt+JL
Insert	Insert	Tables	Table or Home	Slides	New Slide ▼	Title and Content	Table icon	Alt+NT
View gridlines	Table Tools Layout	Table	View Gridlines	Alt+JLTG
Textbox, insert	Insert	Text	Text Box	Alt+NX

esthetically pleasing; however, some themes provide better contrast between background and content than others.

To apply a different Theme to a presentation-in-progress, click a button under **Design | Themes**. Hold the mouse pointer over a Theme (without clicking) to activate Live Preview. The Live Preview feature allows you to see how the change affects the slide before actually applying it. Clicking a **Theme** button applies the design to all slides in the presentation.

Views for Organizing and Writing Your Presentation

There are two tabs on the Ribbon where you can select how you view your presentation:

- Click **View | Presentation Views** to choose views that you will use when you prepare your presentation. **View | Master Views** lets you make changes to the format of every slide in the presentation, the handouts, and the notes.
- Click **Slide Show | Start Slide Show** to see what the slides will look like during the actual presentation.

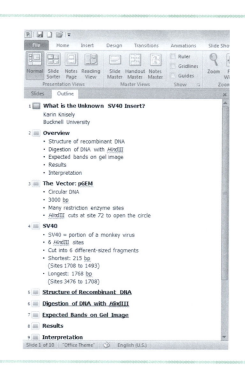

Figure A3.4   The **Outline** tab in **Normal** view helps you structure your presentation. The slide layouts for each slide are chosen by clicking **Home | Slides | Layout**.

The most frequently used views can also be accessed by clicking a button in the lower right corner of the screen (see "Adjust view" in Figure A3.1).

Most of the time, you will be working in **Normal** view. As shown in Figure A3.1, **Normal** view consists of an Overview pane (with two tabs, **Slides** and **Outline**), a Slide pane whose layout and content can be customized, and a Notes pane where you can write notes for your presentation, which will not appear on the slides.

The **Outline** tab on the Overview pane is handy for making an outline of your presentation first (Figure A3.4). As you work on the outline, think about how to present this information using as many visuals and as little text as possible. Then select a slide layout for each main point. For each slide, add the text and the visuals to the Slide pane. For a preview of the show, click **Slide Show** view.

The **Slide Sorter**, **Slide Master**, and **Handout Master** buttons will be described in the section "Revising and Polishing Presentations."

Slide Layouts

Presentations in biology usually follow the Introduction-Body-Closing format. In the **Introduction**, the speaker guides the audience from information that is generally familiar to them to new information, which is the focus (**Body**) of the talk. The take-home message for the audience and any acknowledgments are given in the **Closing**.

Title slide

The default first slide in a new presentation is the Title Slide (see Figure A3.1). Follow the instructions in the placeholders to add text or add custom content (Figure A3.5). The same content will appear on the **Slides** tab of the Overview pane to the left.

The Title Slide is going to set the tone for your presentation. A catchy title and interesting visuals are likely to capture your audience's attention. If possible, allude to the main benefit your audience will gain from listening to your presentation.

Adding slides

To insert a slide after the Title Slide, click **Home | Slides | New Slide ▼** and choose a slide layout (Figure A3.6). The name of the dialog box is the Theme

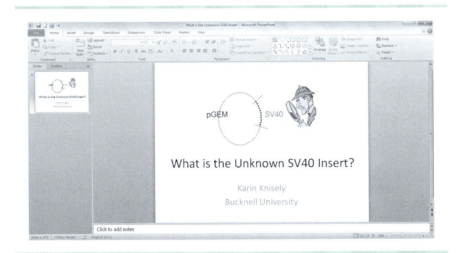

Figure A3.5   Title slide modified by repositioning placeholders and adding graphics and clipart.

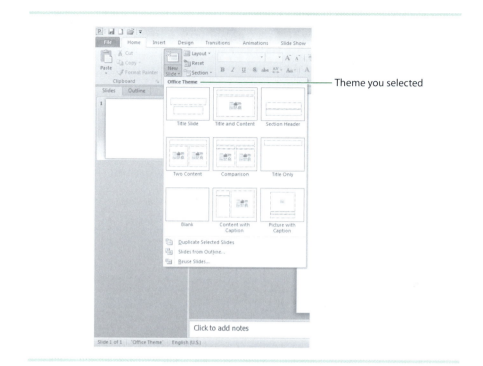

Theme you selected

Figure A3.6   Basic slide layouts are selected by clicking **Home | Slides | New Slide ▼**. There are options for selecting existing slides or outlines from other compatible files.

you selected for your presentation. The Theme for a Blank Presentation is **Office Theme**.

Basic layouts. The basic slide layout options allow you to arrange text and content (pictures, graphs, tables, movies, and graphics) in one or two columns. The placeholders, however, can be modified to suit your needs (see the next section on "Custom Layouts"). Alternatively, if you decide that a different layout would work better after you've already added content to a slide, click **Home | Slides | Layout** and select a different layout.

Custom layouts. It's easy to convert one of the basic layouts to a custom layout by **modifying placeholders** and **inserting visuals** (pictures, shapes, SmartArt Graphics, graphs, text boxes, and media clips).
    Placeholders can be modified as follows:

- To **delete** one, click the dotted border of the placeholder and hit the **Delete** key.

- To **resize** one, click inside the placeholder to display selection handles. Position the mouse pointer over a handle in one of the corners to display a two-headed arrow. Hold down the left mouse button and drag the border to the desired size. Release the mouse button.

- To **move** one, position the cursor over the placeholder to display crossed arrows, hold down the left mouse button, drag the image to the desired location, and release the mouse button.

- To **format text** within one, select the text, click **Home | Font**, and change the typeface, font size, color, and so on. See the section "Formatting Text" on the next page.

Visuals can be inserted by clicking the **Insert** tab and selecting one of the more common options:

- Tables group   Table
- Images group   Picture
- Illustrations group   Shapes, SmartArt, and charts (graphs)
- Links group   Hyperlinks
- Text group   Text boxes
- Media group   Video and audio files

Graphs, tables, and images from Word, Excel, or other files can also be inserted into a PowerPoint slide using Copy and Paste. Visuals don't require a placeholder; they can be added anywhere on the slide and resized or repositioned as needed. Different kinds of visuals will be described in detail later.

Deleting slides

To delete a slide,

- In **Normal** view, click the slide you want to delete on the **Slide** tab on the Overview pane and hit **Delete** or **Backspace**.

- In **Slide Sorter** view, click the slide you want to delete and hit **Delete** or **Backspace**.

Warning: PowerPoint does not ask you if you're sure that you want to delete a slide. If you delete one by accident, however, just click **Undo** on the **Quick Access Toolbar** to get it back.

The last slide

A professional slide show has a definitive ending. End the show with an acknowledgments slide or add a slide that invites the audience to ask questions.

Navigating among Slides in Normal View

There are several ways to move from one slide to another when preparing a presentation in **Normal** view (Figure A3.7):

- Click the **Page Up** or **Page Down** key on the keyboard.
- Use the vertical scroll bar on the right side of the Slide pane.
    - Drag the scroll box inside the scroll bar up or down to move backward or forward through the presentation.
    - Click the Previous Slide or Next Slide arrow buttons.
- Use the Overview pane on the left side of the screen. Click the slide you want to display. *Note*: There are two tabs on the Overview pane: Outline (text) and Slides (thumbnails). Choose the view you prefer.

Formatting Text

Just as in Microsoft Word, you can apply formatting such as typeface, font size, bold, italics, underline, alignment, and color to selected words or lines in a single slide by using the buttons in the **Home | Font** group. A good rule

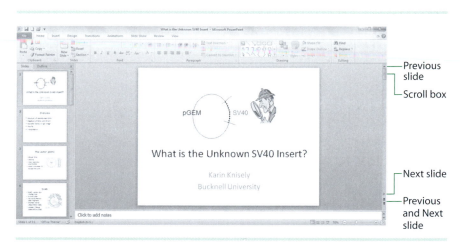

Figure A3.7   Ways to navigate among slides in **Normal** view.

of thumb is to use no more than two fonts or colors, and to use them only for emphasis. For example, boldfaced, italicized, larger sized, and colored text stands out against the default text.

A bulleted list works well if you want to give an overview of your talk first (Figure A3.8A). Each bulleted item should be a short phrase with keywords or key points, not full sentences. The phrases themselves should be informative but interesting for your audience. A good rule of thumb is to use no more than 6 words per bullet and no more than 6 bulleted items on one slide.

To subscript or superscript text, select the character(s) to be formatted, click the **Home | Font** dialog box launcher (the diagonal arrow in the right corner of the group label), and click the appropriate box under **Effects** in the **Font** dialog box.

To insert symbols (Greek letters, mathematical symbols, Wingdings, etc.), click **Insert | Symbols | Symbol**, and look for the desired symbol. For a uniform look, select the same font in the **Symbol** dialog window as the text font in your PowerPoint presentation. It is not possible to make shortcut keys or to program symbols in AutoCorrect in PowerPoint 2010. If the symbols occur frequently in your presentation, however, you can copy and paste them to save yourself some keystrokes.

The AutoCorrect feature (**File | Options | Proofing | AutoCorrect Options**) has only limited usefulness in PowerPoint 2010. In Word 2010, AutoCorrect can be programmed for symbols, italicizing scientific names of organisms automatically, and replacing long chemical names with a simple keystroke combination. Only the latter trick works in PowerPoint (see the "AutoCorrect" section in Appendix 1).

If you want to make changes that affect the text in every slide in the presentation, read the section on "Tweaking Slide, Handout, and Notes Masters" on p. 293. The default lettering (44 pt for titles and 32 pt for text) is designed to be large enough to be read even in the back of an average-sized classroom, so don't reduce font size to get more information on the slide. Instead, keep the wording simple.

Sample Presentation

The sample PowerPoint presentation we are using here is about a laboratory exercise in which a novel recombinant DNA, made of a plasmid (pGEM) and a fragment of a virus that infects monkeys (SV40), was analyzed by agarose gel electrophoresis. Six different recombinant DNAs are possible, each one with the same plasmid (pGEM), but with a different SV40 insert. Since the SV40 inserts are all different sizes, and since agarose gel electrophoresis separates DNA fragments based on size, the goal of the exercise was to determine which of the six possible SV40 inserts was contained in the unknown recombinant DNA.

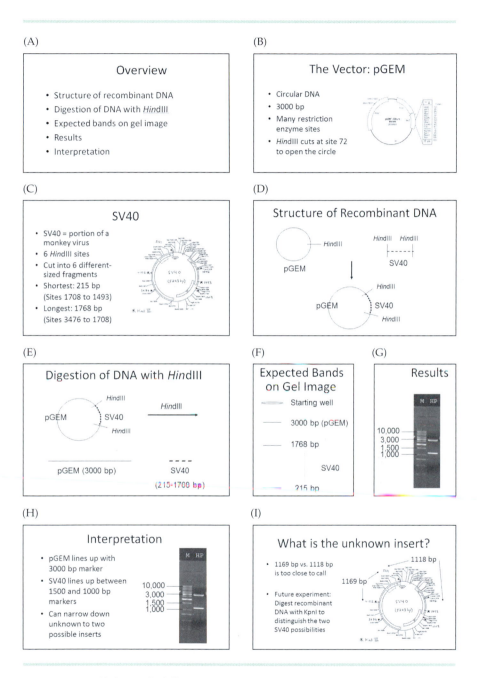

Figure A3.8   Slides with different combinations of text and visuals in a presentation about recombinant DNA analysis using agarose gel electrophoresis.

pGEM and SV40 were joined using a restriction enzyme called *Hind*III. In the lab exercise, the recombinant DNA was digested (cut apart) with *Hind*III, separating the DNA into its two components. The sample was then run on a gel, with the SV40 fragment migrating farther than the pGEM fragment because of its smaller size. By comparing the distance migrated by the SV40 fragment with a standard marker DNA, the identity of the SV40 fragment (based on its size) can be determined.

Figures A3.8B, C, H, and I are slides from the sample presentation, which have **both text and a visual**. Figures A3.8D, E, F, and G are slides with **just a visual**, and Figure A3.5 (the title slide) and Figure A3.8A are **text-only** slides. All of the slides were created from one of the basic layouts (**Home** | **Slides** | **New Slide** ▼) or by modifying the placeholders and inserting an illustration (see the "Adding Slides" section on p. 270). The section on "Line Drawings" on p. 279 describes how to make the simple graphics shown in Figures A3.7D and E.

Visuals

The visuals are presented in the order of the command buttons on the **Insert** tab.

Tables

PowerPoint allows you to insert large and small tables on a slide by selecting any of the layouts with the word "Content" or "Comparison" from the New Slide drop-down menu (see Figure A3.6). In the center of the "Click to add text" placeholder, click the **Insert Table** icon (Figure A3.9) and enter the number of rows and columns in the **Insert Table** dialog box. It's also possible to add a table to a **Blank** layout with **Insert** | **Tables** | **Table**.

As in Word 2010, the table format can be modified with the **Table Tools Design** and **Layout** tabs, which only appear when the insertion pointer is inside the table (Figure A3.10). By convention, tables in scientific papers do not have vertical lines to separate the columns, and horizontal lines are used only to separate the table caption from the column headings, the headings from the data, and the data from any footnotes. To format a table like this, select all cells in the table and click **Design** | **Table Styles**. Scroll through the styles until you reach the "Light Shading 1" series. This series has the appropriate horizontal lines and no vertical lines preferred in scientific papers. For a presentation, the colored rows are appropriate and help the audience follow the arrangement of data in the table.

Navigating in tables. In addition to the mouse, you can use keyboard keys to navigate within a table:

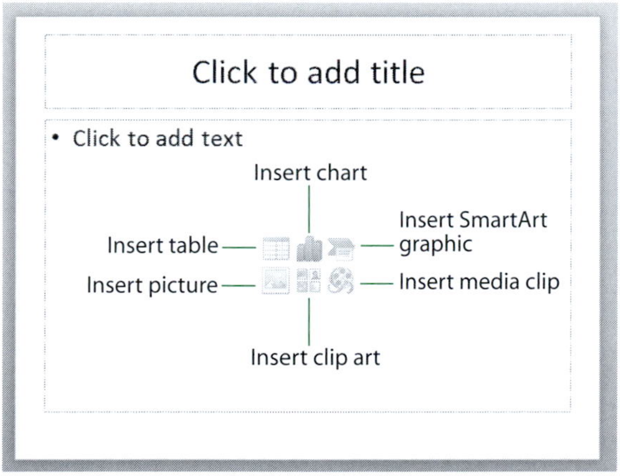

Figure A3.9   To insert a large table, click **Home | Slides | New Slide ▼ | Table and Content** and then click the **Insert Table** icon in the center of the slide.

- To jump from one cell to an adjacent one, use the arrow keys.
- To move forward across the row, use the **Tab** key. *Note:* If you press **Tab** when you are in the last cell of the table, another row will be added to the table.
- To align text on a tab stop in a table cell, press **Ctrl+Tab**.

(A)

(B)

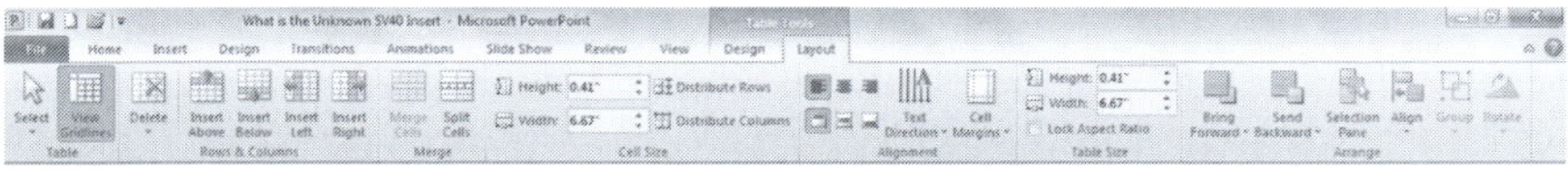

Figure A3.10   Table formatting tabs appear when you click inside a table. (A) The **Design** tab is used primarily to select a style and modify shading and borders. (B) The **Layout** tab is used to insert or delete rows and columns, split or merge cells, adjust cell and table size, and align text within cells. Select **View Gridlines** to make it easier to view individual cells when entering data.

Inserting and deleting rows and columns. To insert a new column or row, position the insertion pointer in a cell adjacent to where you want to insert a new column or row. Select **Table Tools Layout | Rows & Columns** and click on one of the Insert options (**Insert Above**, **Insert Below**, **Insert Left**, or **Insert Right**).

To delete a cell, column, row, or the whole table, position the insertion pointer in the compartment you want to delete. Click the down arrow on the **Delete** button in the **Rows & Columns** group and make your selection.

Merging or splitting cells. To merge two or more cells, select the cells and click **Table Tools Layout | Merge | Merge Cells**. To split a cell into two or more cells, select the cell and click **Table Tools Layout | Merge | Split Cells**. Specify how you want to split the cell: horizontally (enter number of rows) or vertically (enter number of columns).

Changing column width or row height. To change the width of a column, position the cursor on one of the vertical lines so that ◄╫► appears. Then hold down the left mouse button and drag the line to make the column the desired width.

To change the height of a row, position the cursor on one of the horizontal lines so that ╪ appears. Then hold down the left mouse button and drag the line to make the row the desired height.

Viewing gridlines. It is easier to enter data in a table when the gridlines are displayed. Position the insertion pointer inside the table and click **Table Tools Layout | Table | View Gridlines**. Gridlines are not printed.

Formatting text within tables. You can format the text in each cell just as you would format running text. For example, to center the column headings or to put them in boldface, select the top row and click the **Center** or **Bold** button on the **Home** tab. To align numbers, select the entire column and click the **Align Text Right** button. These same options are in the **Table Tools Layout | Alignment** group.

To format all the cells, click the top left cell, hold down the **Shift** key, and click the bottom right cell. To format selected cells, do the same, except hold down the **Ctrl** key.

Pictures

To insert a picture, click **Insert | Images | Picture**. While this function is not new, the ability to edit the picture in PowerPoint is. When you click the picture, the Picture Tools Format tab appears. The buttons on this tab make it possible to remove the background, adjust brightness and contrast, apply artistic effects, and add various borders. Because high resolution images can increase the size of a PowerPoint presentation significantly, compress the pictures if the file will be printed, posted to the Web, or shared via email (**Picture Tools Format | Adjust | Compress Pictures**).

Line drawings

To make line drawings or add simple graphics to your document, click **Insert | Illustrations | Shapes** (Figure A3.11). To insert a line, for example, click a line to display crosshairs. Position the mouse pointer on the slide where you want the line to start, hold down the left mouse button, drag to where you want the line to end, and release the mouse button. To make perfectly horizontal or vertical lines, hold down the **Shift** key while holding down the left mouse button and drag. In general, when it's important that a shape retain perfect proportions (e.g., circle not oval; square not rectangle), hold down the **Shift** key while drawing or resizing the shape with the mouse.

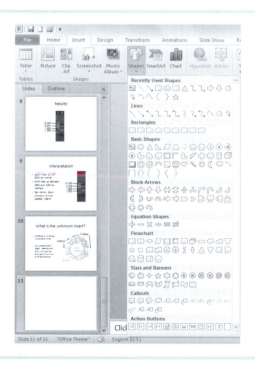

Figure A3.11   Select shapes for simple graphics from the **Insert | Illustrations | Shapes** drop-down menu.

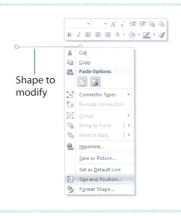

Shape to modify

Figure A3.12 Right-click a shape to size and position the shape precisely (**Size and Position**) or to modify attributes such as fill, weight, color, and style (**Format Shape**).

To change the angle or length of the line, click the line and then position the mouse pointer over one of the end points of the line to display a two-headed arrow. Hold down the left mouse button, drag to where you want the line to end, and release the mouse button.

To change the location of the line, click the line and then position the mouse pointer over the line to display crossed arrows. Hold down the left mouse button, drag the line to the desired location, and release the mouse button. For more precise positioning, right-click the line and select **Size and Position** from the drop-down menu (Figure A3.12).

To format the line, right-click it and select **Format Shape** (see Figure A3.12). This dialog box gives you options for changing the appearance of

Align

Group

Figure A3.13 To format a shape, click it to open the **Drawing Tools Format** tab. Right-clicking the shape opens a dialog box with additional options.

the line. Many of the same formatting commands are located on the **Draw-ing Tools Format** tab that appears when you select the line (Figure A3.13).

To add text to your drawing, click **Insert | Text | Text Box**. Position the mouse pointer where you want the text box to appear, hold down the left mouse button, drag to enlarge the text box to the size you think you'll need (you can resize it later), and release the mouse button. Type text inside the box. Font face and size can be changed by selecting the text and using the buttons in the **Home | Font** group. To format the text box itself, click the text box to display the selection handles on the border. Right-click the text box to open a drop-down menu similar to the one in Figure A3.12 or use the buttons on the **Drawing Tools Format** tab.

If several objects make up a graphic, then it may make sense to group them as a unit. Grouping allows you to copy, move, or format all the objects in the graphic at one time. To group individual objects into one unit, click each object while holding down the **Shift** key. Release the **Shift** key and then click **Group** under **Drawing Tools Format | Arrange**. One set of selec-tion handles surrounds the entire unit when the individual objects are grouped (Figure A3.14). To ungroup, simply select the group and click **Group | Ungroup**.

The **Align** button is handy for precisely lining up objects (shapes as well as text boxes). Click each object that you want to line up while holding down the **Shift** key. Release the **Shift** key, click **Align** under **Drawing Tools For-mat | Arrange**, and select one of the alignment options. Another useful fea-ture for arranging objects on a page is **Align | View Gridlines**. This com-mand puts a non-printing grid on your page. Click **Align | Grid Settings** to adjust the spacing between the gridlines or to snap objects to the grid. Check the **Snap objects to grid** box if you want to position objects on grid-lines. Do not check this box if you prefer to position the objects freely. Finally, to move objects just a fraction of a millimeter from their current position, click the object, hold down the **Ctrl** key, and use the arrow keys to nudge the object exactly where you want it on the slide.

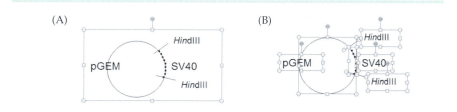

Figure A3.14   Individual graphics grouped (A) and ungrouped (B).

SmartArt Graphics

SmartArt Graphics was first introduced in Office 2007. SmartArt is a collection of diagrams that can be used instead of a bulleted list to show processes, stages or steps, hierarchies, concepts, and other relationships. You might want to use them when you give an overview of your presentation, to describe a procedure, to allude to future projects resulting from the current work, and other situations in which a visual aid can get your point across more effectively than words alone.

To add SmartArt to a slide, choose a slide layout with a Content placeholder (see Figure A3.6) and then click the **Insert SmartArt Graphic** icon (see Figure A3.9) to open the **Choose a SmartArt Graphic** gallery (Figure A3.15). Choose a category on the left or just scroll through all the graphics in the center pane. When you find a graphic that you might like to use, click it to display a detailed description in the right pane. Click **OK** to insert it on the slide. Another way to add SmartArt is by clicking **Insert | Illustrations | SmartArt**.

After you have inserted the SmartArt graphic on a slide, fill the components with text and modify the format if desired. To format the entire graphic, click inside the placeholder. To format individual components of the graphic, click the component. Either way, the **SmartArt Tools** formatting tabs appear on the Ribbon (Figure A3.16). Use the buttons to change the layout and color scheme of both the components and the text. The Live Preview feature allows you to see how the change affects the graphic before actually applying it. Simply hold the mouse pointer over a button on the

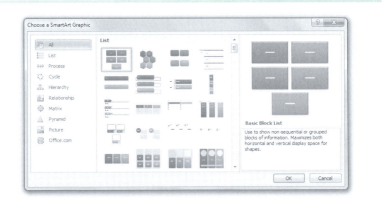

Figure A3.15   A SmartArt Graphic can be used instead of a bulleted list to show relationships.

(A)

(B)

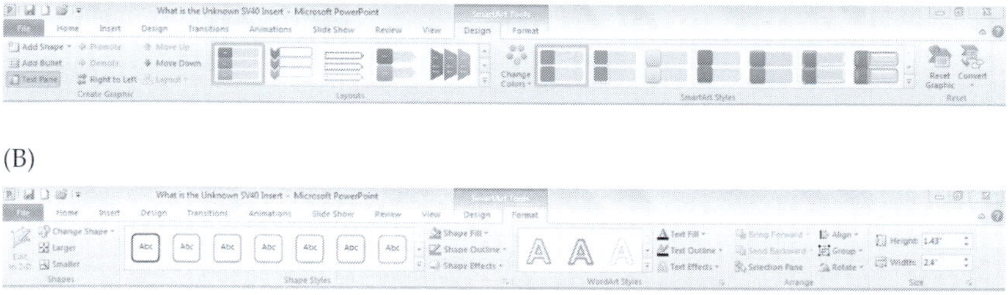

Figure A3.16   SmartArt Tools formatting tabs appear when you click the graphic. (A) The **Design** tab is used to add components and modify the layout and color scheme. (B) The **Format** tab is used to change the shape, size, color, text, and arrangement of the graphic components.

Ribbon (without clicking) to activate Live Preview. When you've decided on a style or layout, click it.

To convert a bulleted list to SmartArt, right-click anywhere within the list and select **Convert to SmartArt**. Then choose an appropriate graphic.

Graphs (Charts)

If you are already familiar with making graphs in Microsoft Excel, then the easiest way to add a graph to a PowerPoint presentation is to copy and paste as follows:

- Make a graph in Excel (see Appendix 2).
- Single-click the Chart Area to activate the graph and then copy it.
- In PowerPoint, click **Home | Slides | New Slide ▼** and choose a layout suitable for the graph.
- Click **Paste**. Resize and reposition the graph as needed.

Hyperlinks

There may be times during a presentation when you'd like to refer to another slide, access an Excel spreadsheet or Word document, or link to an Internet address. These connections are made by inserting hyperlinks into your presentation. Two ways to insert hyperlinks are:

- Using the **Action** button
- Using the **Insert Hyperlink** button

For both methods, you can attach the hyperlink to text (which will be underlined to identify the link) or more subtly to an illustration such as a picture, clip art, shape, or SmartArt graphic. In **Normal** view, type text or insert an object using **Insert | Illustrations**. To attach the hyperlink to text, simply right-click the text and click **Hyperlinks** on the menu. To attach the hyperlink to a visual element (picture, clip art, etc.), right-click the element and then click **Hyperlinks**. Alternatively, click the text or visual and then **Insert | Links | Hyperlink**.

Action button. Click **Insert | Links | Action** to open the **Action Settings** dialog box (Figure A3.17). Use the Mouse Click tab if you would like to activate the link with a mouse click; if you prefer to activate the link by moving the mouse over the object, use the Mouse Over tab. From the **Hyperlink to** drop-down menu, select one of the options to which to link the active slide: a slide in the current presentation, a URL to connect to an Internet address, a different PowerPoint presentation, or a file with a different format (such as an Excel spreadsheet or a Word document). Use the Insert Hyperlink button method if you need to browse for the URL address or copy and paste it from another location.

Insert Hyperlink button. Click **Insert | Links | Hyperlink** to open the **Insert Hyperlink** dialog box (Figure A3.18A). To link to another file on your computer, click the **Existing File or Web Page** button on the left side of the

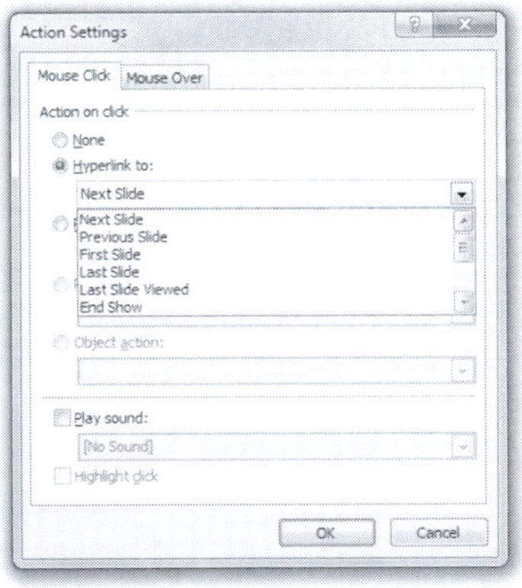

Figure A3.17 To open the **Action Settings** dialog box, click **Insert | Links | Action**. Link the active slide to other slides in the presentation, to a website, or to other Microsoft Office files (e.g., Word, Excel, or PowerPoint) by making a selection on the drop-down menu.

(A)

Click this button to link to file or website

Enter URL here

(B)

Click this button to link to another slide in the current presentation

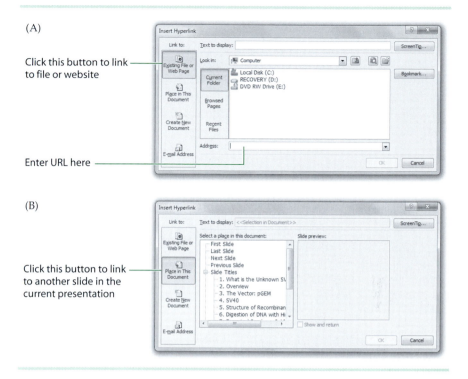

Figure A3.18   To open the **Insert Hyperlink** dialog box, click **Insert | Links | Hyperlink**. (A) Click the **Existing File or Web Page** button to link the active slide to a website or to another Microsoft Office file. (B) Click the **Place in This Document** button to link to another slide in the current presentation.

box. To browse for the file to link, click the **Current Folder** or the **Recent Files** button in the **Look in** area, select the desired file, and click **OK**.

To link to a Web page, click the **Existing File or Web Page** button on the left side of the box (Figure A3.18A). If you have visited the website recently, click the **Browsed Pages** button in the **Look in** area and select the desired URL. Alternatively, launch a Web browser such as Internet Explorer, Firefox, or Safari, find the desired website, and copy and paste the URL into the **Address** box. Click **OK**.

To link to another page in your current presentation, click the **Place in This Document** button on the left side of the **Insert Hyperlink** dialog box (Figure A3.18B). Select the slide to link from the **Select a place in this document** area and click **OK**.

A clever way to return to the slide in the presentation where you inserted the hyperlink is to add an "invisible" return hyperlink. For example, let's say when I reach Slide 8, where I present the results of my study (see Figure A3.8G), I want to refer back to the structure of the recombinant DNA shown on Slide 5 (see Figure A3.8D). After showing Slide 5, I want to return to Slide 8 and continue my presentation. To make a return link on Slide 5, I would position a shape with no line or color somewhere on the slide where I can easily find it, and then link that shape to Slide 8. To add an invisible hyperlink:

- In **Normal** view, insert a shape using **Insert | Illustrations | Shapes**. Select a two-dimensional shape like a rectangle, oval, star, or arrow. Enlarge the shape and position it next to text or in a corner where you will be able to find it easily.

- To make the shape invisible, right-click it and select **Format Shape** from the drop-down menu. Under **Fill**, select **No Fill**. Under **Line Color**, select **No Line**. Then click **Close**. Although you can't see the shape on your slide, you know you've found it when the mouse pointer displays crossed arrows.

- To link the now invisible shape to another slide in the presentation, click the shape and then **Insert | Links | Action** or **Insert | Links | Hyperlink**.

- Follow the instructions given in the sections "Action button" and "Insert Hyperlink button" to make the link.

Preview hyperlinks. To try out the hyperlinks, switch to presentation mode by clicking the **Slide Show** button in the lower right corner of the screen or **Slide Show | Start Slide Show**. Links don't work in **Normal** view. The linked website, page, or slide will open when you mouse click or mouse over a hyperlink.

Multimedia files

Movie and video clips, music, and sounds may enhance your presentation. To embed a video file that you have saved on your computer, click **Insert | Media | Video | Video from file**, browse your computer, and select from various file formats. To embed a YouTube video (Moyea PowerPoint E-Learning Center, accessed on 26 July 2012 from http://www.dvd-ppt-slideshow.com/blog/):

1. Go to www.youtube.com and find the desired video.
2. Below the link, click **Share** and then **Embed**. Check the **Use old embed code** box and then click a resolution.
3. Copy the embed code.

4. In **Normal** view in PowerPoint, click **Insert | Media | Video | Video from Website** and paste the embed code into the text box. Click **Insert** to exit the dialog box. A black box will be inserted on the slide, which you can size and reposition as needed.
5. In presentation mode, click the play button on the icon to start the video.

Warning!

1. Not all embedded videos will play in PowerPoint due to copyright restrictions. In that case, copy the URL into a slide layout with a Content Placeholder (see Figure A3.6). In presentation mode, click the link to the website.
2. The embed code method described above does not work with the **Insert Media Clip** button on Content Placeholder slides (see Figure A3.9).
3. Some videos require the installation of additional software such as QuickTime for the clip to work. If you are having problems playing multimedia files in PowerPoint, ask someone in computer services for help.
4. Embedded media files may drastically increase the size of a PowerPoint file. When possible, compress media by clicking **File | Info | Compress Media** and choosing one of the three options:
   a. **Presentation quality** results in about the same video and audio quality, but makes the file smaller
   b. **Internet quality** results in a reduction in quality comparable to streamed video
   c. **Low quality** is sufficient for attaching the PowerPoint to an e-mail

   (Microsoft Office, http://office.microsoft.com/en-us/ powerpoint-help/compress-your-media-files-HA010382163.aspx, accessed 26 July 2012)

Saving and Printing Presentations

To save a presentation for the first time, click **File | Save**. Do not wait until you are finished making the entire presentation to save the file. Save it after you have made the title slide.

1. Click **File | Save** (or **Save As**). The **Save As** dialog box appears.

2. Click a link in the breadcrumb trail in the box at the top or use the navigation pane on the left to locate the folder in which you want to save the presentation.

3. To create a new folder, click the **New Folder** button. The **New Folder** dialog box appears. Give the folder a short, descriptive name. Click **OK**.

4. Choose a descriptive name for the file.

PowerPoint 2010 automatically saves the file as a "Presentation," with the .pptx extension. If you would like to share a PowerPoint 2010 presentation with someone who has an earlier version of PowerPoint (97–2003) and who has not installed the Compatibility Pack (see the section on "Good Housekeeping" in Appendix 1), click **File | Save As** and next to **Save as type,** select **PowerPoint 97-2003 Presentation**.

If you need to print colored slides on transparencies, because the presentation room doesn't have projection equipment for a PowerPoint presentation,

- Specify the desired size and orientation of the slides with **Design | Page setup | Page setup**
- On the print menu (**File | Print | Full Page Slides**), change the default **Grayscale** setting to **Color** (Figure A3.19A).

As a speaker, you may wish to print out an outline of your presentation or notes pages to refer to when you deliver the presentation. Both of these options are available on the center panel upon clicking **File | Print | Full Page Slides** (Figure A3.19B). When selecting **Notes Pages**, a thumbnail of the slide is printed along with the speaker notes below it. The speaker notes are comparable to notecards in that they help you (the presenter) remember the important points to make about each slide. By sticking to the script, you are likely to stay within your time limit and avoid getting sidetracked.

There is only one option for printing notes pages from PowerPoint: Each printed page consists of one slide with its corresponding notes page below it. If you do not write notes for every slide, this format is wasteful because you end up with very little content on a pile of paper. A more economical alternative is to combine multiple slides with their speaker notes on one sheet of paper. To do so, click **File | Save & Send | File Types | Create Handouts**. Select a page layout and then click **OK** in the **Send to Microsoft Office Word** dialog box (Figure A3.20). A new Word document is created, which consists of a 3-column table with slide number, thumbnail of the slide, and speaker notes in the three columns, respectively. The row height can be adjusted to fit up to 5 slides with corresponding speaker notes on one page.

Figure A3.19 **File | Print** provides options for printing slides, handouts, and notes pages.

By printing your notes pages from Word instead of PowerPoint, you can significantly reduce your paper usage.

Your audience might appreciate having a handout with thumbnails of the slides for note-taking purposes. To print handouts, click **File | Print | Full Page Slides | Handouts** (see Figure A3.19B). Between 1 and 9 slide thumbnails can be printed on one page (the more thumbnails, the smaller

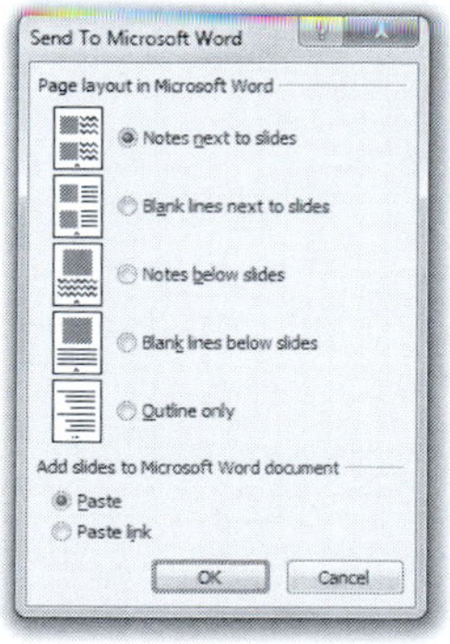

Figure A3.20 To save paper when printing speaker notes, export the PowerPoint slides along with their notes pages and print from Word.

each thumbnail). The 3-slides-per-page option comes with lines next to each thumbnail to facilitate note-taking. When printing on a black-and-white printer, select **Pure Black and White** to avoid printing out the background of the slides (see Figure A3.19A).

Revising and Polishing Presentations

As described previously, the most frequently used views can also be accessed by clicking a button in the lower right corner of the screen (see "Adjust view" in Figure A3.1).

Slide Sorter view is useful for evaluating the overall appearance of a presentation because it displays thumbnails of all of the slides in the presentation, complete with slide transitions (Figure A3.21). You cannot edit the content of an individual slide in **Slide Sorter** view, but you can rearrange the slides. To select a slide to move, copy, or delete, single-click it. To edit the content of the slide, double-click it to return to **Normal** view. Slides can also be rearranged in the Overview pane in **Normal** view.

Before following the instructions below, click **View | Presentation Views | Slide Sorter** to change to **Slide Sorter** view.

Moving slides

To move a slide in Slide Sorter view, position the mouse pointer on the slide you want to move, hold down the left mouse button and drag the vertical line to a new location between slides, and release the mouse button. The slide now appears at its new position. You can also move a slide using

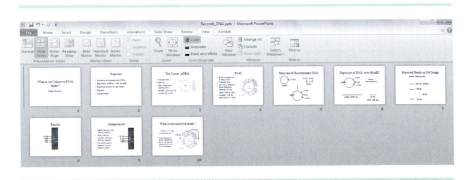

Figure A3.21   **Slide sorter** view makes it easy to rearrange slides in a presentation.

the **Cut** and **Paste** buttons on the **Home** tab, the Ctrl+X and Ctrl+V keyboard shortcuts, or the right mouse button.

To move a group of adjacent slides, click the first slide in the group, hold down **Shift** on the keyboard, and click the last slide in the group. All slides will be marked with a highlighted border. Hold down the left mouse button, drag the vertical line to a new insertion point between slides, and release the mouse button. The group of slides appears at its new position. As with a single slide, you can also use Cut and Paste.

To move a group of nonadjacent slides, follow the procedure for adjacent slides, but hold down **Ctrl** instead of **Shift** on the keyboard. The slides appear in the same consecutive order at the new location.

Adding and deleting slides

To add a new slide in Slide Sorter view, click the insertion point between two slides. Then click **Home | Slides | New Slide ▼** and choose a slide layout. Double-click the new slide to add content.

To delete an unwanted slide, single-click the slide and press **Delete** on the keyboard or click **Home | Slides | Delete**. To correct the deletion, click **Quick Access Toolbar | Undo**.

Copying slides

To copy a slide in Slide Sorter view, use the **Copy** and **Paste** buttons on the **Home** tab, the Ctrl+C and Ctrl+V keyboard shortcuts, or the right mouse button. To insert a duplicate slide immediately after the original, click the original and then **Home | Slides | New Slide ▼ | Duplicate Selected Slides**. To change the layout on the duplicate, simply single-click the duplicate slide, click **Home | Slides | Layout**, and select a new layout for the slide. This method was used to make Figure A3.8H from Figure A3.8G.

Spell check

Don't let an otherwise interesting and well-organized presentation be ruined by typos on the slides. As with Microsoft Word, PowerPoint gives you visual indicators to alert you to possible spelling mistakes. Words that are possibly misspelled are underlined with a wavy red line. To deal with this word, right-click it and correct the mistake, tell PowerPoint to ignore it, or, if the word is a correctly spelled technical term, add it to the dictionary. In contrast to Word, PowerPoint does not flag grammatical errors.

Applying a theme

In an effective presentation, the slides are simple, legible, and organized log-ically. Although there is nothing wrong with black-and-white slides (using the default Office Theme), most people prefer color. PowerPoint offers a variety of color-coordinated design templates (called "themes"). To see how your presentation would look with a different theme, click the **More** arrow in the **Design | Themes** group and hold the mouse over each of the but-tons. The Live Preview feature lets you see how the theme affects slide con-tent before committing to the new theme. When you find a theme that you like, clicking the button applies the theme to all slides in the presentation. It may actually be better to select a theme before rather than after writing your presentation to avoid conflicts between slide content and the design later on. If you later decide to go back to the default theme for a Blank Pre-sentation and it's too late to use the **Undo** command, click **Design | Themes | Office Theme** (the first theme in the gallery).

To change the slide background without applying a theme, click **Design | Background | Background Styles**. If you select one of the available styles, the text color will be adjusted automatically (e.g. black text on a white back-ground will be converted to white text on a dark background). If you choose a background color from the **Format Background** dialog box (**Design | Background** dialog box launcher), however, text color does not change and you may have to adjust it manually.

Animations and slide transitions

Animation affects the movement of individual objects on a slide during a slide show. Without animation, each visual element or line of text is present when the slide appears. However, when you would like to give the audi-ence time to focus on and process one point at a time (for example, when listing objectives or summarizing conclusions), it is advantageous to use animations.

Animations are selected on the **Animations** tab (Figure A3.22A). In **Nor-mal** view, click the **More** arrow next to the animations and hold the mouse over the buttons to preview the animation. Adjust the direction and sequence of the animation effect with the **Effect Options** button. To copy an animation to another object on the slide (or another slide), click the ani-mated object, then select **Animations | Advanced Animation | Animation Painter**, and then click the new object to animate.

Transitions affect the way slides appear in a slide show. Without slide transitions, each slide stays on the screen until you click the left mouse but-ton, press **Enter**, or press the space bar on the keyboard to advance to the next slide. The slides in the presentation are displayed sequentially unless

(A)

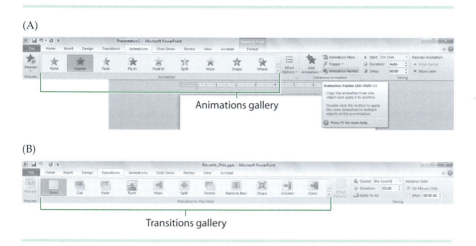

Animations gallery

(B)

Transitions gallery

Figure A3.22 (A) Animations can be applied to individual objects on a slide. (B) Slide transitions affect how slides change from one to the next.

you press the **Esc** key to end the show, click a button on the Slide Show toolbar in the lower left corner of the screen, or right-click and choose an option from the menu. Transitions are selected on the **Transitions** tab (Figure A3.22B).

For a tutorial on using animations and transitions in PowerPoint 2010, see http://office.microsoft.com/en-us/powerpoint-help/animations-and-transitions-RZ102809184.aspx?CTT=1. If you decide to use animations and slide transitions, be careful not to overuse them. *You want your audience to be listening to you, not be distracted by the special effects!*

Tweaking Slide, Handout, and Notes Masters

The Slide Master controls the appearance of the content in every slide *including* the title slide. The Title Master in earlier versions of PowerPoint has been replaced with a Title Style layout within the **Slide Master** in PowerPoint 2010.

You can modify the Slide Master at any time—before, during, or after making the first draft of your presentation. Click **View | Master Views | Slide Master** to open an Overview pane, with the Slide Master thumbnail at the top and the layouts that belong to it beneath it (Figure A3.23). The Slide pane shows the selected thumbnail in **Normal** view. If you want to make changes that affect every slide in the presentation, be sure to modify the Slide Master, not one of the layout slides.

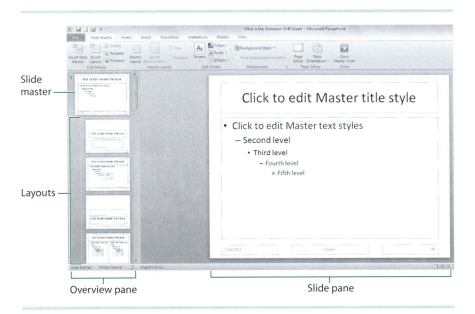

Slide master —

Layouts —

Overview pane

Slide pane

Figure A3.23 Default **Slide Master** (Office Theme).

Text, bullets, and illustrations

To modify the font face, style, size, or color, select the line of text you want to change. Right-click and select **Font** from the drop-down menu. Make your selection and click **OK**. To change a bullet, click anywhere on the bulleted line. Right-click and select **Bullets** from the drop-down menu. Make your selection and click **OK**. To add a logo to the Slide Master, click **Insert | Images** and then make the appropriate selection. **Remember that these changes affect every slide in the presentation.**

Headers and footers

The three boxes at the bottom of the Slide Master allow you to enter the date and time, a custom footer, and the page number. While the default setting is *not* to include this information during a slide show, it may be handy to have it on the printed slides if they are used as handouts (Figure A3.24). To include the footer information:

- Click **View | Presentation Views | Slide Master**.

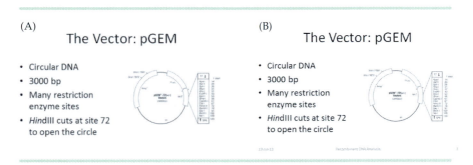

Figure A3.24 Footers on slides can be turned off (A) or on (B) by checking the appropriate boxes on the **Slide** tab in the **Header and Footer** dialog box.

- Click **Insert | Text | Header and Footer** to open the **Header and Footer** dialog box (Figure A3.25).
- On the **Slide** tab, click the **Date and Time**, **Slide Number**, and/or **Footer** check boxes.
- Click **Apply to All** to print this information on every slide.
- Click **File | Print** and select **Settings | Full Page Slides** from the Print options panel.
- Make any other changes to the settings and then click **OK**.

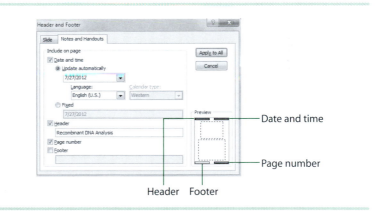

Figure A3.25 The **Header and Footer** dialog box allows you to print the date and time, a page number, and a custom header and footer on Notes and Handouts pages. Similar information can be included (or not) on the slides by making the appropriate selections on the **Slide** tab.

Handouts and notes pages

Checking the footer boxes on the **Slide** tab of the **Header and Footer** dialog box causes footers to be displayed on *slides*, but not on handouts or notes pages. To print headers and/or footers on these pages:

- Click **View | Presentation Views | Handout Master**.
- Click **Insert | Text | Header and Footer** to open the **Header and Footer** dialog box.
- On the **Notes and Handouts** tab, click the relevant check boxes (see Figure A3.25). Date and time are displayed in the upper right corner and the page number in the lower right corner. A custom header and footer can be entered in the respective boxes if desired.
- Click **Apply to All** to print the selected information on every page.
- Click **File | Print** and select **Settings | Full Page Slides | Handouts** or **Notes Pages** from the Print options panel.
- Make any other changes to the settings and then click **OK**.

Delivering Presentations

A PowerPoint presentation can be "delivered" electronically by attaching it to an e-mail or posting it to a website, but most presentations are delivered in person. Table A3.2 gives some presentation tips. See Chapter 8 for detailed instructions on planning and delivering oral presentations.

Resources in the presentation room

When you are invited to give a talk, find out what audio-visual equipment will be available in the room. If you only have a chalkboard, just print out the speaker notes. If you'll have access to an overhead projector and screen, print out the slides and make transparencies. These printing options are explained in the section on "Saving and Printing Presentations" on pp. 287–290.

To deliver a PowerPoint slide show, you'll need a computer and projection equipment as well as a screen on which to display the slides. If the presentation room does not have this equipment, ask your host if he/she can provide it or, if you are affiliated with a college or university, you may be able to borrow a laptop computer and projector for your presentation.

There are several ways to run a PowerPoint presentation:

- From a Web server

TABLE A3.2 Presentation tips	
Do	**Don't**
Keep the wording simple	Write every word you're going to say on the slide
Make the text large and legible	Use backgrounds that make text and images hard to read
Strive for a consistent look (use the same font and format for each slide)	Use distracting animations and slide transitions, or sound effects
Include visuals that complement and support the auditory information	Include tables when you can show the trend better with a graph
Allow enough time for each slide	Talk endlessly without referring to a visual
	Rush through the slides without mentioning the take-home message about each one

- From a flash drive or CD on a computer at the presentation site
- From the hard drive of your laptop computer

To run a presentation from a Web server, first find out if you have access to the Internet at the presentation site. If so, save your file to a remote server, such as Google Drive, SkyDrive, or Dropbox. Similarly, if your college or university has networked computers, you can prepare your presentation on a networked computer in your room, for example, save the file in your Netspace, and then access the file from another networked computer in the room where you will hold your talk.

If the presentation site has a computer, but it is not hooked up to the Internet, save your presentation file on a flash drive or CD and carry it with you. Ask your host if PowerPoint is installed on the presentation room's computer. If so, just plug the flash drive into the USB port, start PowerPoint, open your presentation file, and click the **Slide Show** button in the lower right corner of the screen.

If you have a laptop computer, make sure PowerPoint is installed on it, and then save your presentation file on the hard drive. At the presentation site, connect your laptop to the projector, start PowerPoint, open the file, and click the **Slide Show** button in the lower right corner of the screen to begin your presentation.

Navigating among slides during a slide show

To start a slide show *from the first slide*, click the **F5** key or **Slide Show | Start Slide Show | From Beginning**. Clicking the **Slide Show** button in the lower right corner of the screen or **Shift + F5** starts your presentation *from the current slide*. It is possible to specify a time for the slides to advance automatically (**Transitions | Timing | Advance Slide After**), but it is more common to use the keyboard, the mouse, or a wireless presenter to advance to the next slide.

The following keystrokes allow you to navigate among slides when you deliver your presentation:

- To advance to the next slide, press **Enter** or the space bar.
- To go back to the previous slide, press **Backspace**.
- To end a slide show, press **Esc**.

To change slides using the mouse:

- Click a button on the **Slide Show** toolbar in the lower left corner of the screen (Figure A3.26).
- Click the left mouse button to advance to the next slide.
- Click the right mouse button to display the main menu of navigation options.

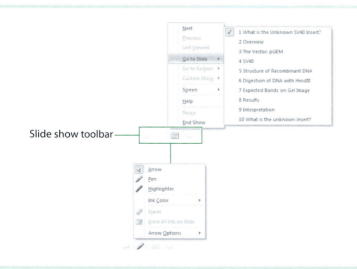

Slide show toolbar

Figure A3.26   Options for navigating among slides during a slide show. The **Slide Show** toolbar has arrows to go forward and backward, a pointer button, and a slide button for navigation options.

To turn the mouse into a laser pointer:

- In **Slide Show** view, hold down the **Ctrl** key while holding down the left mouse button.
- Choose a color for the pointer by clicking **Slide Show | Set Up | Set Up Slide Show**.

A wireless presenter allows you to change slides from anywhere in the room. This device usually consists of a USB receiver that you plug into the computer's USB port, and the remote control, which has buttons for moving forward and backward, pausing, or stopping the slide show. The remote control may also have a built-in laser pointer.

To write on the slides:

- In **Slide Show** view, right-click and select **Pointer Options | Pen** (or **Highlighter**). See Figure A3.26.
- Hold down the left mouse button and write.

To erase "ink" on the slides, right-click and select **Erase All Ink on Slide**.

Bibliography

Databases and scholarly search engines

Bakkalbasi N, Bauer K, Glover J, Wang L. c2006. Three options for citation tracking: Google Scholar, Scopus and Web of Science. Biomedical Digital Libraries [Internet]. [cited 2012 Aug 20]; 3:7. Available from: http://www.bio-diglib.com/content/3/1/7

Falagas ME, Pitsouni EI, Malietzi GA, Pappas G. c2008. Comparison of PubMed, Scopus, Web of Science, and Google Scholar: strengths and weaknesses. The FASEB Journal 22: 338–342

Giustini D, Barsky E. c2005. A look at Google Scholar, PubMed, and Scirus: comparisons and recommendations. JCHLA / JABSC [Internet]. [cited 2012 Aug 20] 26: 85–89. Available from: pubs.chla-absc.ca/doi/pdf/ 10.5596/c05-030

Recommended Search Engines: Tables of Features [Internet]. c2008. Berkeley (CA): Regents of the University of California. [updated 2008 Jul 13; cited 2012 Aug 20]. Available from: http://www.lib.berkeley.edu/ TeachingLib/Guides/Internet/SearchEngines.html

Sadeh T. c2006. Google Scholar Versus Metasearch Systems. High Energy Physics Libraries Webzine [Internet]. [cited 2012 Aug 20]; Issue 12. Available from: http://library.cern.ch/HEPLW/12/papers/1/

Shultz M. c2007. Comparing test searches in PubMed and Google Scholar. J Med Libr Assoc [Internet]. [cited 2012 Aug 20]; 95(4): 442–445. Available from: http://www.pubmedcentral.nih.gov/articlerender.fcgi? artid=2000776

Evaluating web pages

Alexander J, Tate M. c1996-2005. Evaluate Web Pages [Internet]. Chester (PA): Widener University; [cited 2012 Aug 20]. Available from: http://www.widener.edu/about/campus_resources/wolfgram_library/ evaluate/default.aspx

Evaluating Web Pages: Techniques to Apply & Questions to Ask [Internet]. c2008. Berkeley (CA): Regents of the University of California. [updated 2008 Jul 13; cited 2012 Aug 20]. Available from: http://www.lib.berkeley. edu/TeachingLib/Guides/Internet/Evaluate.html

Olin & Uris Libraries: Critically Analyzing Information Sources [Internet]. c2010. Ithaca (NY): Cornell University Library. [updated 2011 Apr 5; cited 2012 Aug 20]. Available from: http://olinuris.library.cornell.edu/ref/research/skill26.htm

Purdue Online Writing Lab (OWL) [Internet]. c1995-2012. West Lafayette (IN): The Writing Lab & The OWL at Purdue and Purdue University. [updated 2012 Mar 12; cited 2012 Aug 20]. Available from: http://owl.english.purdue.edu/owl/resource/553/01/

Internet sources, citing

Patrias K. 2007-. Citing medicine: the NLM style guide for authors, editors, and publishers [Internet]. 2nd ed. Wendling DL, technical editor. Bethesda (MD): National Library of Medicine (US); [updated 2011 Sep 15; cited 2012 Oct 30]. Available from: http://www.nlm.nih.gov/citingmedicine

Microsoft Office

Gookin D. 2010. Microsoft Word 2010 for Dummies. Hoboken, NJ: Wiley Publishing, Inc. 408 p.

Harvey G. 2010. Microsoft Office Excel 2010 for Dummies. Hoboken, NJ: Wiley Publishing, Inc. 408 p.

How to Build Powerful PowerPoint Presentations. 2007. Mission (KS): CompuMaster.

LabWrite Resources. c2004. Graphing Resources: Using Error Bars in your Graph [Internet]. Raleigh (NC): North Carolina State University; [cited 2012 Aug 20]. Available from: http://www.ncsu.edu/labwrite/res/res-homepage.htm

Miscellaneous

Light RJ. 2001. Making the most of college: Students speak their minds. Cambridge, MA: Harvard University Press. 242 p.

Martin B. 2011. Doing good things better. Ed (Sweden): Irene Publishing. [cited 2012 Nov 1]. Available free online at: http://www.bmartin.cc/pubs/11gt/

Meece M. 2012. A User's Guide to Finding Storage Space in the Cloud. [Internet] The New York Times. [publ. 2012 May 16; accessed 2012 Aug 20]. Available from: http://www.nytimes.com/2012/05/17/technology/personaltech/a-computer-users-guide-to-cloud-storage.html?pagewanted=all

Wynn ES. The Best Ways To Backup Your Files And Photos — Advice And Tips. HubPages, Inc. [updated 2012 Jan 5; accessed 2012 Aug 20]. Available from: http://earlswynn.hubpages.com/hub/What-is-the-best-way-to-backup-your-files-and—photos——-advice-and-tips-_

Oral presentations

D'Arcy J. 1998. Technically Speaking: A Guide for Communicating Complex Information. Columbus: Battelle Press. 270 p.

Fegert F, Hergenröder F, Mechelke G, Rosum K. 2002. Projektarbeit: Theorie und Praxis. Stuttgart: Landesinstitut für Erziehung und Unterricht Stuttgart. 226 p.

Hailman JP, Strier KB. 2006. Planning, proposing, and presenting science effectively: A Guide for Graduate Students and Researchers in the Behavioral Sciences and Biology, 2nd ed. Cambridge: Cambridge University Press. 248 p.

Plagiarism

Frick, Ted. 2004. Understanding Plagiarism [Internet]. Bloomington (IN): Indiana University, School of Education; [updated 2004 Sept 18; cited 2012 Aug 20]. Available from: http://www.indiana.edu/~tedfrick/plagiarism/

Posters

Alley M. c1997. Design of Scientific Posters [Internet]. The Pennsylvania State University: State College (PA); [last updated April 2008; cited 2012 Aug 20]. Available from: http://www.writing.engr.psu.edu/posters.html

Cothran K. c2009. Research Poster Design, Construction, and Printing. [cited 2012 Aug 20]. Available from: http://www.clemson.edu/centers-institutes/crca/documents/templates/poster-design-presentation.pdf

Hess GR, Tosney K, Liegel L. c2006. Creating Effective Poster Presentations [Internet]. NC State University: Raleigh (NC) [cited 2012 Aug 20]. Available from:
http://www.ncsu.edu/project/posters/NewSite/CreatePosterText.html

Ritchison G. c2009. BIO 801 Scientific Literature and Writing Poster Presentations [Internet]. Eastern Kentucky University: Richmond (KY). [cited 2012 Aug 20]. Available from:
http://people.eku.edu/ritchisong/posterpres.html

Stocks SD, Zimmerman JK. c2004. SC Life Publication Series: Guidelines for creating a poster for a professional meeting [Internet]. SC LIFE Clemson University: Clemson (SC). [cited 2012 Aug 20]. Available from:
http://www.clemson.edu/cafls/sclife/documents/how_to_create_a_poster_ur.pdf

Woolsey JD. 1989. Combating poster fatigue: How to use visual grammar and analysis to effect better visual communications. Trends in Neuroscience 12(9): 325–332.

Reading and writing about biology

Gillen CM. 2007. Reading Primary Literature. San Francisco: Pearson/Benjamin Cummings. 44 p.

Hofmann AH. 2010. Scientific Writing and Communication: Papers, Proposals, and Presentations. New York: Oxford University Press. 682 p.

Lannon JM, Gurak LJ. 2011. Technical Communication, 12th ed. White Plains (NY): Longman. 784 p.

McMillan VE. 2012. Writing Papers in the Biological Sciences, 5th ed. Boston: Bedford/St. Martin's. 256 p.

Palmer-Stone D. 2001. How to Read University Texts or Journal Articles [Internet]. Victoria (BC), Canada: University of Victoria, Counselling Services; [updated 2009 Dec 17; cited 2012 Aug 20]. Available from: http://www.coun.uvic.ca/learning/reading-skills/texts.html

Pechenik JA. 2012. A Short Guide to Writing about Biology, 8th ed. Upper Saddle River (NJ): Pearson Education Addison-Wesley. 288 p.

VanAlstyne JS. 2005. Professional and technical writing strategies, 6th ed. New York, NY: Pearson Longman. 752 p.

Revision

Cook Counseling Center – Virginia Tech. 2009. Proofreading [Internet]. Virginia Tech: Blacksburg, VA [cited 2012 Nov 1]. Available from: http://www.ucc.vt.edu/stdysk/proofing.html

Corbett PB. 2011. The Reader's Lament [Internet]. The New York Times: New York. [cited 2012 Oct 31]. Available at: http://afterdeadline.blogs.nytimes.com/2011/10/04/the-readers-lament/

Every B. c2006-2012. My 15 Best Proofreading Tips [Internet]. BioMedical Editor [cited 2012 Nov 1]. Available from: http://www.biomedical editor.com/proofreading-tips.html

LR Communication Systems, Inc. c1999. Proofreading and editing tips: A compilation of advice from experienced proofreaders and editors [Internet]. Berkeley Heights, NJ [cited 2012 Oct 31]. Available from: http://www.lrcom.com/tips/proofreading_editing.htm

The Writing Center at UNC Chapel Hill. c2010-2012. Revising Drafts [Internet]. University of North Carolina: Chapel Hill, NC [cited 2012 Oct 31]. Available from: http://writingcenter.unc.edu/handouts/revising-drafts/

Scholarly communication issues

Scholarly Communication [Internet]. 2007. Urbana-Champaign (IL): University of Illinois at Urbana-Champaign. [updated 2009 Aug 13; cited 2012 Aug 20]. Available from: http://www.library.illinois.edu/scholcomm/

Suber P. 2004. Open Access Overview [Internet]. Richmond (IN): Earlham College; [revised 2012 June 18; cited 2012 Aug 20]. Available from: http://www.earlham.edu/~peters/fos/overview.htm

Scientific style and format

Council of Science Editors [Internet]. c2008. Reston (VA): Council Science Editors, Inc.; [cited 2012 Aug 20]. Available from: http://www.council sci-enceeditors.org/

Council of Science Editors, Style Manual Committee. 2006. Scientific style and format: the CSE manual for authors, editors, and publishers. 7th ed. Reston (VA): The Council. 680 p.

Peterson SM. 1999. Council of Biology Editors guidelines No. 2: Editing science graphs. Reston, VA: Council of Biology Editors. 34 p.

Peterson SM, Eastwood S. 1999. Council of Biology Editors guidelines No. 1: Posters and poster sessions. Reston, VA: Council of Biology Editors. 15 p.

Statistics

Moore DS, Notz WI, Fligner MA. 2011. The Basic Practice of Statistics, 6th ed. New York: W.H. Freeman and Company. 745 p.

Writing guides

Hacker D, Sommers N. 2012. A pocket style manual, 6th ed. Boston: Bedford/St. Martin's. 304 p.

Lunsford AA. 2010. Easy Writer, 4th ed. Boston: Bedford/St. Martin's. 352 p.

Lunsford AA. 2013. The Everyday Writer, 5th ed. Boston: Bedford/St. Martin's. 608 p.

Lunsford AA. 2011. The New St. Martin's Handbook, 7th ed. Boston: Bedford/St. Martin's. 992 p.

INDEX

The designation *t* following a page number indicates the information will be found in a table, and the designation *f* following a page number indicates the information will be found in a figure.